FIXED EFFECTS ANALYSIS OF VARIANCE

Probability and Mathematical Statistics

A Series of Monographs and Textbooks

S. Vajda. Probabilistic Programming. 1972

Sheldon M. Ross. Introduction to Probability Models. 1972

Robert B. Ash. Real Analysis and Probability. 1972

V. V. Fedorov. Theory of Optimal Experiments. 1972

K. V. Mardia. Statistics of Directional Data. 1972

H. Dym and H. P. McKean. Fourier Series and Integrals. 1972

Tatsuo Kawata. Fourier Analysis in Probability Theory. 1972

Fritz Oberhettinger. Fourier Transforms of Distributions and Their Inverses: A Collection of Tables. 1973

Paul Erdös and Joel Spencer. Probabilistic Methods in Combinatorics. 1973

K. Sarkadi and I. Vincze. Mathematical Methods of Statistical Quality Control. 1973

Michael R. Anderberg. Cluster Analysis for Applications. 1973

W. Hengartner and R. Theodorescu. Concentration Functions. 1973

Kai Lai Chung. A Course in Probability Theory, Second Edition. 1974

L. H. Koopmans. The Spectral Analysis of Time Series. 1974

L. E. Maistrov. Probability Theory: A Historical Sketch. 1974

William F. Stout. Almost Sure Convergence. 1974

E. J. McShane. Stochastic Calculus and Stochastic Models. 1974

Z. Govindarajulu. Sequential Statistical Procedures. 1975

Robert B. Ash and Melvin F. Gardner. Topics in Stochastic Processes. 1975

Avner Friedman, Stochastic Differential Equations and Applications, Volume 1, 1975; Volume 2. 1975

Roger Cuppens. Decomposition of Multivariate Probabilities. 1975

Eugene Lukacs. Stochastic Convergence, Second Edition. 1975

H. Dym and H. P. McKean. Gaussian Processes, Function Theory, and the Inverse Spectral Problem. 1976

N. C. Giri. Multivariate Statistical Inference. 1977

Lloyd Fisher and John McDonald. Fixed Effects Analysis of Variance. 1978

In Preparation

Sidney C. Port and Charles J. Stone. Brownian Motion and Classical Potential Theory.

Fixed Effects Analysis of Variance

LLOYD FISHER
JOHN McDONALD
Department of Biostatistics
University of Washington
Seattle, Washington

ACADEMIC PRESS New York San Francisco London 1978
A Subsidiary of Harcourt Brace Jovanovich, Publishers

ACADEMIC PRESS, INC.
111 Fifth Avenue, New York, New York 10003

United Kingdom Edition published by
ACADEMIC PRESS, INC. (LONDON) LTD.
24/28 Oval Road, London NW1 7DX

Library of Congress Cataloging in Publication Data

Fisher, Lloyd, Date
Fixed effects analysis of variance.

(Probability and mathematical statistics
;)

1. Analysis of variance. I. McDonald, John, Date joint author. II. Title.
QA279.F57 519.5'35 77-92238
ISBN 0-12-257350-1

PRINTED IN THE UNITED STATES OF AMERICA

TO

Ginny and the Christian fellowship of
Gerald, Ian, Carl, Busso, John, Ed, Art, and Ross

LF

TO

my family and friends for putting up with the perpetual student

JMcD

CONTENTS

PREFACE

This book is designed as a reference and also as a textbook for a one-quarter course for statistics students at the upper-division undergraduate or the beginning graduate level. It was written because in the authors' opinion there is no book suitable for a one-quarter course on a moderately advanced theoretical level. Most theoretical textbooks are too long and are designed for a one-year course. When the books are used for a one-quarter course, the course is almost exclusively theory without examples, which often has a detrimental effect on the motivation of the students. Textbooks designed for a one-quarter length course tend to be on a theoretical level that is too low. Thus, this book has been designed to bridge the one-quarter gap between books such as those by Gunther and Scheffé, for example. Care has been taken to use examples from the literature in order to give students the emotional as well as mental realization that the techniques are indeed used for day-to-day data analysis.

Prerequisites are a background in probability and statistics at roughly the level of Hogg and Craig. It is useful if the probability and statistics course has covered the concepts of hypothesis testing, p values, confidence intervals, and moment generating functions. The reader should also have a background in linear algebra. At the University of Washington this background, although required, has usually been rusted somewhat by the time interval between the linear algebra course and the analysis of variance course. Also, the linear algebra courses often do not cover certain useful topics, such as projection operators and quadratic forms, that are crucial for the analysis of variance. For this reason, an appendix reviewing results of linear algebra is included. There is particular emphasis on projection operators, a topic slighted in some linear algebra books, and acquaintance with the results of Appendix 1 will save the reader many hours which might be spent thumbing through the diverse linear algebra literature. The appendix, which gives this background, is at a higher level than is needed for most of the text. It is felt that when this book is used as a textbook the instructor may need to adjust the lectures to the level and material of the linear algebra course taught at the particular school.

The final chapter in the book is a short introduction to multiple regression. This was included since the general linear model is available, waiting to be used at the end of the text. It is felt that this chapter will enable the reader to make the connection between analysis of variance and multiple regression techniques as they are studied.

Results of Appendix 1 are prefaced by A; for example, Theorem 1 of the appendix is referred to in the text as Theorem A1.

This book is not intended to be a terminal point in the education of the reader. It is hoped that this book will enable one to tackle the more advanced texts by Scheffé, *The Analysis of Variance*, Searle, *Linear Models*, and Kempthorne, *Design and Analysis of Experiments*, as well as books of a more applied bent, such as Winer, *Statistical Principles in Experimental Design.*

The scope of this book has been quite modest as befits a book written for a one-quarter course. Some extensions of the theory pointing the way to generalization and to other techniques are given in the problems at the ends of the chapters.

ACKNOWLEDGMENTS

The authors wish to acknowledge the kind cooperation and permission of the many authors and publishers of examples cited in the text. Thanks are due to The Society for Child Development, Inc. for giving permission to use their copyrighted material that appeared in *Child Development* **46**, 258–262; **47**, 237–241, 532–534 as used in Example 2.2 and Problem 6.14.

Thanks are also due to the American Journal of Nursing Company for permission to use their copyrighted material from *Nursing Research* that appears in Example 2.1. We also wish to thank Marcel Dekker, Inc., and Professors Odeh, Owen, and Birnbaum for allowing the use of the tables given in Appendix 2.

Students who worked through versions of the text for two consecutive years are thanked for their kind comments and sufferance. Thanks are also due to Ms. Janice Alcorn and Ms. Cindy Bush. Finally, one author (LF) would like to thank his family, who were patient during many hours of work on the text at home.

1 INTRODUCTION

The subject matter we will consider is extremely beautiful. The mathematical theory of the fixed effects analysis of variance is elegant and can give considerable insight into much of statistical theory after the material has been mastered. Also, when applied to appropriate data, it gives a very concise and elegant method of analyzing quite complex situations. Unfortunately, between the elegant theory and elegant application there is a fair amount of detail and unaesthetic algebraic manipulation. If one keeps sight of the beginning and the end results of the process, however, the algebraic manipulations should not prove too much of a nuisance.[1]

The term "analysis of variance," and indeed much of the fundamental work in the subject, is the work of Sir Ronald A. Fisher. In the words of Fisher:

> . . . the analysis of variance, that is, by the separation of variance ascribable to one group of causes from the variance ascribable to other groups.[2]

This, then, is the thrust of the subject we are going to study: to see which part of variation in data might reasonably be attributed to various causes and which part of the variability seems to be due to uncontrollable variation, whether biological, physical measurement error, or otherwise.

[1] Beauty and elegance are in the eye of the beholder, and this view, of course, represents that of the authors of this material.

[2] R. A. Fisher, *Statistical Methods for Research Workers*, 11th ed., revised, p. 211. Hafner, New York, 1950

It is interesting to consider Fisher's background, which enabled him to develop the analysis of variance. Mahalanobis says of him:

> Love of mathematics dominated his educational career; and he was fortunate in coming under the tuition of a brilliant mathematical teacher, W. N. Roe of Stanmore Park, also well known in England as a Somersetshire cricketer. At Harrow he worked under C. H. P. Mayo and W. N. Roseveare. The peculiar circumstances of the teaching by the latter will be of interest to statisticians familiar with Fisher's geometrical methods. On account of his eyes he was forbidden to work under artificial light, and when he would go to work with Mr. Roseveare in the evenings the instruction was given purely by ear without the use of paper and pencil or any other visual aid. To this early training may be attributed Fisher's tendency to use hypergeometrical representations, and his great power in penetrating problems requiring geometrical intuition. It is, perhaps, a consequence of this that throughout life, his solutions have been singularly independent of symbolism. He does not usually attempt to write down the analysis until the problem is solved in his mind, and sometimes, he confesses, after the key to the solution has been forgotten. I have already mentioned that Fisher was not acquainted with Students' work when he wrote his 1914 paper on the exact distribution of the correlation coefficient. Here he introduced for the first time the brilliant technique of representing a sample of size n by a point in a space of n dimensions. Such representation has proved extremely useful in subsequent work not only in the theory of distribution, but also in other fields of statistical theory such as the work of J. Neyman and E. S. Pearson.[3]

In line with these comments we find that the mathematics needed to understand the analysis of variance involves working in n-dimensional Euclidean space. While the reader of this text is expected to have had some background in linear algebra, some review may prove necessary (see Appendix 1).

The usefulness of Fisher's mathematical background resulted from the fact that he was basically concerned with experimentation and applications.

> It is a common misapprehension among mathematicians that new statistical methods can be developed solely by the process of pure reasoning without contact with sorted numerical data. It was part of Fisher's strength that although an early mathematician, he held no brief for this belief. He was himself a skilled and accurate computer,

[3] P. C. Mahalanobis, Professor Ronald Almer Fisher, *Biometrics* **20**, 238–251 (June 1964).

> and did not hesitate to embark on the analysis of any body of data that interested him. The rapid advancement of experimental design and analysis owe much to this.[4]

Fisher's contribution to the analysis of variance is summarized as follows:

> Perhaps no technique is more characteristic of Fisher, more closely associated with his name, or more central to his pattern of thought than the analysis of variance. Arising almost incidentally during his efforts to make sense of agricultural experiments whose designs were inadequate for sound inference, this rapidly became the standard way of analyzing results from the next generation of experiments.[5]

In this text we shall present the theoretical ideas and some applications of the analysis of variance. The emphasis will be on understanding the theoretical background behind the analysis of variance. We shall, however, attempt to interject enough examples so the reader will see how the theoretical material is related to applications and will be in a position to pursue the subject further. Near the end of this text, we shall see that the models we are considering are appropriate not only for the analysis of variance but for understanding many other multivariate statistical topics. Thus, we shall touch upon topics known as multiple regression analysis, and certain other aspects of multivariate statistics such as multiple and partial correlation coefficients.

[4] F. Yates, Appreciations of R. A. Fisher, *Biometrics* **20**, No. 2, 312–313 (June 1964).

[5] D. J. Finney, Sir Ronald Fisher's contributions to biometric statistics, *Biometrics* **20**, 322–329 (June 1964).

2 THE t-TEST

Many of the ideas we need for the analysis of more complex situations are contained in the familiar t-test. Let us review this. Random variables $X_1, \ldots, X_n$ are a random sample if they are independent and identically distributed (i.i.d.) random variables. A random variable X is normally distributed with mean μ and positive variance σ^2, denoted by $N(\mu, \sigma^2)$, if X has the probability density function

$$f(x) = \frac{1}{\sqrt{2\pi\sigma^2}} \exp\left[-\frac{1}{2}\left(\frac{x-\mu}{\sigma}\right)^2\right].$$

If $X_1, \ldots, X_n$ is a random sample, the sample mean is

$$\bar{X} = \frac{1}{n}\sum_{i=1}^{n} X_i$$

and the sample variance is

$$s^2 = \frac{1}{n-1}\sum_{i=1}^{n}(X_i - \bar{X})^2.$$

In deriving the t-test it is shown that if $X_1, \ldots, X_n$ is a $N(\mu, \sigma^2)$ random sample, then $\bar{X}$ is $N(\mu, \sigma^2/n)$ and $(n-1)s^2/\sigma^2$ has a χ^2_{n-1} (chi-square) distribution with $n-1$ degrees of freedom (df). Further, $\bar{X}$ and s^2 are independent random variables. From this it follows that $\sqrt{n}(\bar{X} - \mu)/\sigma$ is $N(0, 1)$ and independent of $s/\sigma = \sqrt{(n-1)s^2/(n-1)\sigma^2}$ which is the square root of a chi-square distributed random variable divided by its degrees of freedom. The ratio of these two independent random variables $\sqrt{n}(\bar{X} - \mu)/s$ has by definition a t-distribution with $n-1$ df.

Example 2.1 [M. S. Brown and J. T. Hurlock, Preparation of the breast for breastfeeding, *Nursing Research* **24**, No. 6 (1975)][1] One is often interested in differences between treatment effects rather than the treatment effects themselves. Suppose that n pairs of similar individuals or things or experimental materials are selected. To compare two treatments, one treatment is applied to one member of each pair, the other treatment to the other member. Ideally the treatments are assigned at random. The data consists of a random sample of n differences in measurement for the n pairs. The differences $D_i, i = 1, \ldots, n$, are assumed to be independent and identically distributed normal random variables, that is, $D_i \sim N(\mu, \sigma^2)$. Usually, it is desired to test the hypothesis H_0: $\mu = 0$, that is, the mean difference in treatment effects is zero. Such a test is often called a paired t-test or matched t-test.

As an example, consider the study by Brown and Hurlock to evaluate the effectiveness of three commonly suggested methods of preparing the breasts for breastfeeding. The breasts are prepared for breastfeeding prenatally in order to prevent postnatal nipple pain and trauma. Three methods were studied:

(1) nipple friction or rolling, the purpose of which is to "toughen" the epithelium (skin) of the nipple so that it is more impervious to the trauma of sucking;
(2) application of various types of creams to "lubricate" and soften the nipple, making it more supple and hence less susceptible to damage;
(3) the prenatal expression of colostrum (the first milk secreted before or after birth).

Fifty-seven women volunteered to prepare one breast and not the other. The method of preparing one breast and not the other on an individual woman was believed by the investigators to offer a distinct advantage in the control of possibly confounding variables. It meant the vast majority of possibly influential variables (e.g., nutritional status of mother, health conditions, socioeconomic status, race) were held constant since they were true, of course, for both breasts. The breast to be prepared was chosen randomly by a coin toss. Each woman was assigned randomly to the three treatment groups so that each group contained 19 subjects.

Although an investigator visited the subjects to evaluate objective measures of nipple irritation, we shall present only the subject's subjective indications of nipple pain. The mother was asked to fill out a questionnaire immediately after each feeding to rate nipple sensitivity on a scale of 1 to 4 for each

[1] Copyrighted by the American Journal of Nursing Company.

side (from comfortable to painful). The data are presented in Problem 2.22. One-sample t-tests were used to test the null hypothesis of no differences. When the treated and untreated breasts are compared, no significant differences ($\alpha = .05$) in subjective or objective measures of nipple sensitivity or trauma is found is any of the three groups.

Example 2.2 [P. Caplan and M. Kinsbourne, Baby drops the rattle: Asymmetry of duration of grasp by infants, *Child Development* **47**, 532–534 (1976)][2] Consider the Caplan–Kinsbourne study which explored the question, "Does asymmetry in hand use develop, or is it present from birth?" Specifically, they formulated the question as follows: Will children longer attend to and therefore more persistently grasp an object held in the right hand than the left? Twenty-one infants with mean age 2.7 months were tested. There were two parts of the experiment. In the first part, a rattle was placed in one of the baby's hands, and the number of seconds before it was dropped was recorded. For half the subjects the rattle was first placed in the left hand and for half the right. Eight trials were run, alternating hands. In the data analysis, trial 1 was compared with trial 2, etc., in order to minimize the likelihood of changes in mood, cognitive set, or body state from one trial to its matched pair. When the mean numbers of seconds the right and left hands, respectively, held the rattle were compared, a significantly greater right-hand persistence appeared, $R = 61.88$ (mean for the right hand), $L = 40.71$ (mean for the left hand), paired $t = 2.686$, $p < .05$. This is developmentally the earliest reported instance of asymmetry of normal dexterity and/or preference. In the second part of the experiments, two identical rattles were used, and one was placed in each hand simultaneously. The first hand to drop a rattle and the length of time for which grasp persisted were noted. Four trials were run. Part two yielded a right-hand mean of 44.58 sec and a left-hand mean of 35.64 sec, paired $t = 1.38$, not significant.

Example 2.3 [F. Bednarek and D. Roloff, Treatment of apnea of prematurity with aminophylline, *Pediatrics* **58**, No. 3 (Sept. 1976)] A paired t-test is often used when the two treatments are applied to the same experimental unit sequentially. For example, a measurement may be taken before any "treatment" is applied and once again after the treatment.

As an example, consider the study by Bednarek and Roloff on the effectiveness of a specific drug (aminophylline) in the prevention and/or treatment of apnea (absence of spontaneous breathing) of prematurity. Thirteen premature infants who had six or more apneic episodes on each of two consecutive 8-hr shifts were admitted to the study. The responses of the individual patients are seen in Table 2.1. The responses are recorded as apneic episodes

[2]

TABLE 2.1

Response to Aminophylline (Apneic Episodes per Hour)

Dose (mg/kg)	Toxicity[a]	24 hr before treatment	Hours after treatment									Recurrence of apnea (hr)
			8	16	24	32	40	48	56	64	72	
3.8	IT	1.71	.13	.13	.25	.00	.00	.00	.00	.00	.00	48
2.4	TAB	1.25	1.25	.88	.38	.25	.00	.25	.25	.63	.25	—
6.1	—	2.13	.50	1.38	1.13	2.50	2.38	1.50	1.63	1.13	2.50	32
6.0	—	1.29	1.25	.13	.13	.00	.00	.00	.00	.00	.00	24
2.8	—	1.58	.00	.25	.13	.13	.25	.25	.13	.25	.25	—
4.2	—	4.00	2.50	2.63	1.25	1.25	.88	.75	.00	.25	.00	—
3.8	—	1.42	1.75	1.38	.38	.00	.13	.00	.00	.00	.00	32
4.3	IT	1.08	.75	.50	1.00	.88	1.13	.50	1.25	1.13	2.00	1
5.0	—	1.83	1.88	1.25	.00	.25	.50	.50	.25	.50	.50	16
3.3	IT	.67	1.37	.75	1.00	2.38	1.63	.75	.75	.86	.13	32
6.4	—	1.13	2.50	.00	.00	.00	.00	.00	.00	.13	.00	—
3.8	—	2.71	1.25	2.38	2.63	1.36	.63	.38	.25	.38	.50	1
1.7	—	1.96	2.25	1.13	1.50	1.00	.50	.13	.50	.00	.00	8
Mean 4.2		1.72	1.34	.98	.75	.77	.62	.38	.39	.40	.47	22
SD 1.4		±.86	±.83	±.83	±.77	±.89	±.73	±.52	±.52	±.42	±.82	16

[a] IT, intermittent tachycardia (>180 beats/min); TAB, transient anal bleeding.

per hour in order to compare each 8-hr shift with the 24-hr period prior to the treatment. The response for each 8-hr interval when compared to the 24-hr pretreatment period was significant ($p < .01$, paired t-test) for all intervals except the first.

The key to understanding the theory behind the t-test will be a geometric understanding. The random sample $X_1, \ldots, X_n$ will be considered as a random vector

$$\mathbf{X} = \begin{pmatrix} X_1 \\ \vdots \\ X_n \end{pmatrix}$$

taking values in R^n, n-dimensional Euclidean space, a vector space. We will use the usual inner product.

Appendix 1 contains an informal review of linear algebra and vector space theory which stresses a geometric understanding of the material. The appendix also emphasizes the concept of a coordinate system, how a vector is expressed relative to a given ordered basis or coordinate system, and how to change from one coordinate system to another. If the reader feels insecure in this area, it is best to review the appendix before proceeding further.

If we observe data $X_1, \ldots, X_n$, it is natural to record it as

$$\mathbf{X} = \begin{pmatrix} X_1 \\ \vdots \\ X_n \end{pmatrix},$$

that is, the value of the ith observed random variable occupies the ith slot of the n-tuple $\mathbf{X}$. It is often convenient to write a vector as the ordered n-tuple of its coordinates relative to a given ordered basis. The ith coordinate of a vector relative to the given ordered basis occupies the ith position of the n-tuple. This coordinate representation will be used for all vectors in this text. The point is that to express a vector as an n-tuple the coordinate system being used must be understood. Therefore, our data implicitly is recorded with respect to the $\boldsymbol{\delta}_1, \ldots, \boldsymbol{\delta}_n$ coordinate system defined by the condition that the jth coordinate of $\boldsymbol{\delta}_i$ is 1 if $i = j$ and 0 otherwise; that is,

$$\boldsymbol{\delta}_1 = \begin{pmatrix} 1 \\ 0 \\ 0 \\ \vdots \\ 0 \end{pmatrix}, \quad \boldsymbol{\delta}_2 = \begin{pmatrix} 0 \\ 1 \\ 0 \\ \vdots \\ 0 \end{pmatrix}, \ldots, \quad \boldsymbol{\delta}_n = \begin{pmatrix} 0 \\ 0 \\ \vdots \\ 0 \\ 1 \end{pmatrix}$$

where $\boldsymbol{\delta}_i$ has a 1 in the ith position and 0's elsewhere. This particular basis is called the standard or canonical basis. It is important to note that this is an orthonormal coordinate system. The coordinate system appropriate for recording data will not be, in general, the best coordinate system for examining the distribution theory of a random vector. Therefore, it will often be more convenient theoretically to work in a rotated coordinate system.

DEFINITION 2.1 Let $\mathbf{X}$ be a random vector (in R^n) with each entry having a finite mean and variance. The expectation of a random vector is defined as the vector of expectations. Therefore, the *mean vector* of $\mathbf{X}$ is given by

$$\boldsymbol{\mu} = E(\mathbf{X}) = \begin{pmatrix} E(X_1) \\ \vdots \\ E(X_n) \end{pmatrix}.$$

The *covariance matrix* of a random vector is defined as the matrix of covariances. Therefore, the covariance matrix of $\mathbf{X}$ is given by the n by n symmetric matrix $\text{cov}(\mathbf{X})$ whose ijth entry is $\text{cov}(X_i, X_j)$. The expectation of a random matrix is defined as the matrix of expectations. Therefore, the covariance matrix may be written as $\text{cov}(\mathbf{X}) = E\{[\mathbf{X} - E(\mathbf{X})][\mathbf{X} - E(\mathbf{X})]'\}$. (Note that one form of the Cauchy–Schwarz inequality is $\text{cov}^2(X_i, X_j) \leq \text{var}(X_i)\,\text{var}(X_j)$, so that our assumption that each entry of our random vector has finite mean and variance implies that the covariance of each pair X_i, X_j exists.)

Since our work will depend heavily upon the multivariate normal distribution, sometimes shortened to multinormal or just normal distribution, a considerable digression is in order to examine and review the multinormal distribution. Defining these distributions by means of the joint probability density function is not satisfactory because the joint density function does not exist for some of these distributions. In such a case, we say that the multinormal distribution is singular since the covariance matrix is singular. Instead, we will use a more general definition.

DEFINITION 2.2 An R^n-valued random vector $\mathbf{X}$ is *normally distributed* if for each vector $\mathbf{a}$ in R^n, the random variable $\mathbf{a}'\mathbf{X}$ is a (one-dimensional) normal random variable; that is, every linear function of the n variables $X_1, \ldots, X_n$ has a univariate normal distribution. (*Note*: Distributions degenerate at a point are considered normal with zero variance.)

Now consider some of the consequences of this definition. Since $\boldsymbol{\delta}_i'\mathbf{X} = X_i$, each X_i is univariate normal and thus must have a finite mean and variance. Therefore, the mean vector and covariance matrix of a multinormal random vector exist. From Problem 2.3 a multinormal random vector $\mathbf{X}$ with mean vector $\boldsymbol{\mu}$ and covariance matrix $\mathbf{C}$ has the moment generating function

$$M(\mathbf{t}) = \exp(\mathbf{t}'\boldsymbol{\mu} + \tfrac{1}{2}\mathbf{t}'\mathbf{C}\mathbf{t}). \tag{1}$$

Since the distribution of $\mathbf{X}$ is uniquely determined by the moment generating function (Problem 2.4), the multinormal distribution is completely specified by the mean vector and covariance matrix of the random vector. Therefore, if $\mathbf{X}$ is multinormal with mean vector $\boldsymbol{\mu}$ and covariance matrix $\mathbf{C}$, we denote this by $\mathbf{X} \sim N(\boldsymbol{\mu}, \mathbf{C})$ or by saying that $\mathbf{X}$ has a $N(\boldsymbol{\mu}, \mathbf{C})$ distribution.

When the covariance matrix is nonsingular, the multinormal distribution is called nonsingular and the joint probability density function exists. We now find the form of the density function for multinormal $\mathbf{X}$ when the covariance matrix is nonsingular.

THEOREM 2.1 If the n-dimensional random vector $\mathbf{X}$ has a $N(\boldsymbol{\mu}, \mathbf{C})$ distribution with nonsingular covariance matrix, then $\mathbf{X}$ has probability density function

$$f(\mathbf{x}) = |\mathbf{C}|^{-1/2}(2\pi)^{-n/2} \exp[-\tfrac{1}{2}(\mathbf{x} - \boldsymbol{\mu})'\mathbf{C}^{-1}(\mathbf{x} - \boldsymbol{\mu})], \tag{2}$$

where $|\mathbf{C}|$ is the determinant of $\mathbf{C}$.

Proof We shall show that $f(\mathbf{x})$ is the density for a random vector with moment generating function as given in (1). The fact (Problem 2.4) that a moment generating function uniquely determines the distribution then implies the theorem.

It is extremely important to understand the proof of this theorem as it contains some key ideas which will recur in our work. The coordinate system appropriate for recording data will not in general be the best coordinate system for examining the distribution theory of the random vector. In particular, we shall show that there is an orthonormal coordinate system in R^n such that the function in (2) factors into the product of n terms, each involving only one coordinate.

Since the covariance matrix $\mathbf{C}$ is a real symmetric matrix, the Principal Axis theorem of linear algebra (Theorem A14) shows that there exists an orthogonal matrix $\mathbf{P}$ such that $\mathbf{PCP}' = \mathbf{D}$, where $\mathbf{D}$ is a diagonal matrix, that is,

$$\mathbf{D} = \begin{pmatrix} d_1 & 0 & 0 & \cdot & \cdot & \cdot & 0 \\ 0 & d_2 & 0 & \cdot & \cdot & \cdot & 0 \\ 0 & 0 & d_3 & \cdot & \cdot & \cdot & \cdot \\ \cdot & \cdot & \cdot & & & \cdot & \cdot \\ \cdot & \cdot & \cdot & & & \cdot & \cdot \\ 0 & 0 & \cdot & \cdot & \cdot & \cdot & d_n \end{pmatrix},$$

where $\mathbf{PP}' = \mathbf{P}'\mathbf{P} = \mathbf{I}$, the identity matrix.

We evaluate the integral of $f(\mathbf{x})$ over R^n by changing variables by the linear transformation $\mathbf{z} = \mathbf{P}\mathbf{x}$. We may view this orthogonal linear transformation as equivalent to the process of changing from one orthonormal coordinate system to another. Or equivalently, Theorem A9 shows that the vectors $\mathbf{z}$ and $\mathbf{x}$ represent the coordinates of the same data point but with respect to different orthonormal coordinate systems.

Recall that the Jacobian of a linear transformation with nonsingular square matrix is the determinant of the matrix. Recall also that the determinant of a product is the product of the determinants and that the determinant of a matrix equals the determinant of the transpose of the matrix. Thus, $\mathbf{P}\mathbf{P}' = \mathbf{I}$ implies $|\mathbf{P}\mathbf{P}'| = |\mathbf{P}|\,|\mathbf{P}'| = |\mathbf{P}|^2 = |\mathbf{I}| = 1$. Therefore the determinant of an orthogonal matrix is $+1$ or -1. Thus, the absolute value of $|\mathbf{P}|$ is 1, so that the Jacobian of this one-to-one linear transformation is 1. Hence, premultiplying $\mathbf{z} = \mathbf{P}\mathbf{x}$ by $\mathbf{P}'$ implies $\mathbf{P}'\mathbf{z} = \mathbf{x}$, so that

$$\begin{aligned}\int_{-\infty}^{\infty} \cdots \int_{-\infty}^{\infty} f(\mathbf{x})\, dx_1 \cdots dx_n &= \int_{-\infty}^{\infty} \cdots \int_{-\infty}^{\infty} f(\mathbf{P}'\mathbf{z})\, dz_1 \cdots dz_n \\ &= \int_{-\infty}^{\infty} \cdots \int_{-\infty}^{\infty} |\mathbf{C}|^{-1/2} (2\pi)^{-n/2} \\ &\qquad \times \exp[-\tfrac{1}{2}(\mathbf{P}'\mathbf{z} - \boldsymbol{\mu})'\mathbf{C}^{-1}(\mathbf{P}'\mathbf{z} - \boldsymbol{\mu})]\, dz_1 \cdots dz_n.\end{aligned}$$

The exponent is

$$\begin{aligned}-\tfrac{1}{2}(\mathbf{P}'\mathbf{z} - \boldsymbol{\mu})'\mathbf{C}^{-1}(\mathbf{P}'\mathbf{z} - \boldsymbol{\mu}) &= -\tfrac{1}{2}(\mathbf{P}'\mathbf{z} - \mathbf{P}'\mathbf{P}\boldsymbol{\mu})'\mathbf{C}^{-1}(\mathbf{P}'\mathbf{z} - \mathbf{P}'\mathbf{P}\boldsymbol{\mu}) \\ &= -\tfrac{1}{2}(\mathbf{z} - \mathbf{P}\boldsymbol{\mu})'\mathbf{P}\mathbf{C}^{-1}\mathbf{P}'(\mathbf{z} - \mathbf{P}\boldsymbol{\mu}).\end{aligned}$$

Since $\mathbf{P}\mathbf{C}\mathbf{P}' = \mathbf{D}$ and $\mathbf{P}' = \mathbf{P}^{-1}$, we have $\mathbf{D}^{-1} = \mathbf{P}\mathbf{C}^{-1}\mathbf{P}'$, so the exponent is

$$-\tfrac{1}{2}(\mathbf{z} - \mathbf{P}\boldsymbol{\mu})'\mathbf{D}^{-1}(\mathbf{z} - \mathbf{P}\boldsymbol{\mu}) = -\tfrac{1}{2}\sum_{i=1}^{n} d_i^{-1}(z_i - (\mathbf{P}\boldsymbol{\mu})_i)^2.$$

Further, $\mathbf{C} = \mathbf{P}'\mathbf{D}\mathbf{P}$ implies $|\mathbf{C}| = |\mathbf{P}'|\,|\mathbf{D}|\,|\mathbf{P}| = |\mathbf{D}| = \prod_{i=1}^{n} d_i$. Therefore, making the obvious substitutions, we obtain

$$\begin{aligned}&\int_{-\infty}^{\infty} \cdots \int_{-\infty}^{\infty} f(\mathbf{x})\, dx_1 \cdots dx_n \\ &= \int_{-\infty}^{\infty} \cdots \int_{-\infty}^{\infty} \left[\prod_{i=1}^{n} \frac{1}{\sqrt{2\pi d_i}}\right] \exp\left[-\tfrac{1}{2}\sum_{i=1}^{n} d_i^{-1}(z_i - (\mathbf{P}\boldsymbol{\mu})_i)^2\right] dz_1 \cdots dz_n \\ &= \prod_{i=1}^{n} \int_{-\infty}^{\infty} \frac{1}{\sqrt{2\pi d_i}} \exp\left[-\frac{1}{2d_i}(z_i - (\mathbf{P}\boldsymbol{\mu})_i)^2\right] dz_i = \prod_{i=1}^{n} 1 = 1\end{aligned}$$

since the one-dimensional integrands are normal with mean $(\mathbf{P}\boldsymbol{\mu})_i$ and variance d_i. Thus, $f(\mathbf{x})$ is a density function.

Note that the Z_i are independent normal random variables. If only we had started with this coordinate system! The moment generating function for $\mathbf{Z}$ is easy to find (using Problem 2.2) since the Z_i are independent:

$$M_{\mathbf{Z}}(\mathbf{t}) = E(e^{\mathbf{t}'\mathbf{Z}}) = E\left(\prod_{i=1}^{n} e^{t_i Z_i}\right) = \prod_{i=1}^{n} E(e^{t_i Z_i})$$

$$= \prod_{i=1}^{n} \exp[t_i(\mathbf{P}\boldsymbol{\mu})_i + \tfrac{1}{2}d_i t_i^2] = \exp[\mathbf{t}'(\mathbf{P}\boldsymbol{\mu}) + \tfrac{1}{2}\mathbf{t}'\,\mathbf{D}\mathbf{t}].$$

Now the moment generating function for $\mathbf{X}$, $M_{\mathbf{X}}(\mathbf{t})$, can be expressed as a function of the moment generating function for $\mathbf{Z}$, $M_{\mathbf{Z}}(\mathbf{t})$:

$$\begin{aligned} M_{\mathbf{X}}(\mathbf{t}) &= E(e^{\mathbf{t}'\mathbf{X}}) = E(e^{\mathbf{t}'\mathbf{P}'\mathbf{P}\mathbf{X}}) = E(e^{(\mathbf{P}\mathbf{t})'\mathbf{P}\mathbf{X}}) \\ &= E(e^{(\mathbf{P}\mathbf{t})'\mathbf{Z}}) = M_{\mathbf{Z}}(\mathbf{P}\mathbf{t}) \\ &= \exp[(\mathbf{P}\mathbf{t})'\mathbf{P}\boldsymbol{\mu} + \tfrac{1}{2}(\mathbf{P}\mathbf{t})'\mathbf{D}(\mathbf{P}\mathbf{t})] \\ &= \exp[\mathbf{t}'\boldsymbol{\mu} + \tfrac{1}{2}\mathbf{t}'\mathbf{C}\mathbf{t}] \qquad \text{since} \quad \mathbf{P}'\mathbf{D}\mathbf{P} = \mathbf{C}. \end{aligned}$$

Therefore the random vector $\mathbf{X}$ with probability density (2) is multinormally distributed. □

Returning now to our t-test, if $X_1, \ldots, X_n$ is a $N(\mu, \sigma^2)$ random sample where μ and $\sigma^2 > 0$ are unknown parameters, then the joint density function of $\mathbf{X}$ on R^n is

$$f(\mathbf{x}) = \prod_{i=1}^{n} \frac{1}{\sqrt{2\pi\sigma^2}} \exp\left[-\frac{1}{2}\left(\frac{x_i - \mu}{\sigma}\right)^2\right]$$

$$= |\sigma^2\mathbf{I}|^{-1/2}(2\pi)^{-n/2} \exp[-\tfrac{1}{2}(\mathbf{x} - \mu\mathbf{1})'(\sigma^2\mathbf{I})^{-1}(\mathbf{x} - \mu\mathbf{1})],$$

where $\mathbf{I}$ is the n by n identity matrix and $\mathbf{1}' = (1, 1, \ldots, 1)$ is the vector whose entries are all 1. Therefore, by the previous theorem, we observe that $\mathbf{X} \sim N(\mu\mathbf{1}, \sigma^2\mathbf{I})$.

The majority of our work will concern independent observations where each observation has the same variance. In this case the covariance matrix is of the form $\sigma^2\mathbf{I}$ where σ^2 is the variance of each observation. In Theorem 2.1 we changed coordinate systems so that the covariance matrix of our transformed random vector had a particularly simple form; it was diagonal. A covariance matrix of the form $\sigma^2\mathbf{I}$ is already diagonal (with even simpler form since the diagonal entries are equal). For $\mathbf{C} = \sigma^2\mathbf{I}$, *any* orthogonal transformation leads to the same covariance matrix. (Why?) If $\mathbf{X} \sim N(\mu\mathbf{1}, \sigma^2\mathbf{I})$, does the geometry suggest any particular coordinate system?

The expected value of $\mathbf{X}$ lies in a known direction, namely the $\mathbf{1}$ direction. Let $\mathbf{b}_1 = \mathbf{1}/\|\mathbf{1}\|, \mathbf{b}_2, \ldots, \mathbf{b}_n$ be an orthonormal basis for R^n (generated as in Problem 2.18). By Theorem A.9 the n by n orthogonal matrix $\mathbf{P}$ that transforms from the $\boldsymbol{\delta}_i$ coordinate system to the $\mathbf{b}_i$ coordinate system is expressed in terms of its rows as

$$\mathbf{P} = \begin{pmatrix} \mathbf{b}_1' \\ \mathbf{b}_2' \\ \vdots \\ \mathbf{b}_n' \end{pmatrix}.$$

$\mathbf{Z} = \mathbf{PX}$ represents the coordinates of our data point with respect to the new coordinate system but may be viewed as an orthogonal linear transformation of $\mathbf{X}$.

Since linear transformations of multinormal distributions are multinormal (Problem 2.7), we have $\mathbf{X} \sim N(\mu\mathbf{1}, \sigma^2\mathbf{I})$ implies

$$\mathbf{Z} = \mathbf{PX} \sim N(\mathbf{P}\mu\mathbf{1}, \mathbf{P}\sigma^2\mathbf{IP}') \qquad \text{or} \qquad \mathbf{Z} \sim N(\mu\mathbf{P1}, \sigma^2\mathbf{PP}')$$

or

$$\mathbf{Z} \sim N\left(\mu\begin{pmatrix} \mathbf{b}_1'\mathbf{1} \\ \vdots \\ \mathbf{b}_n'\mathbf{1} \end{pmatrix}, \sigma^2\mathbf{I}\right) \qquad \text{or} \qquad \mathbf{Z} \sim N\left(\mu\begin{pmatrix} \frac{\mathbf{1}'\mathbf{1}}{\|\mathbf{1}\|} \\ 0 \\ \vdots \\ 0 \end{pmatrix}, \sigma^2\mathbf{I}\right).$$

Recall (Problem 2.16) that jointly normally distributed random variables are independent if and only if their covariance is zero. Thus $Z_1 \sim N(\mu\sqrt{n}, \sigma^2)$ and $Z_i \sim N(0, \sigma^2)$ for $i = 2, \ldots, n$ are independent random variables. Intuitively, the $\mathbf{1}$ direction contains the information about μ (unfortunately also involving the unknown σ^2). A test for the null hypothesis $\mu = 0$ is now clear. Recall that:

(1) If $Y_1, \ldots, Y_m$ are independent $N(0, 1)$ random variables, then $W = \sum_{i=1}^{m} Y_i^2$ is a $\chi_m{}^2$ random variable with m df.

(2) If V is a $N(0, 1)$ random variable which is independent of W which is a $\chi_m{}^2$ random variable, then $t = V/\sqrt{W/m}$ is a t random variable with m df.

In our case, $(Z_1 - \sqrt{n}\mu)/\sigma$ is $N(0, 1)$ independent of $W = \sum_{i=2}^{n} (Z_i/\sigma)^2$ which is χ^2 with $n - 1$ df. Thus,

$$t = \frac{(Z_1 - \sqrt{n}\mu)/\sigma}{\sqrt{\sum_{i=2}^{n} (Z_i/\sigma)^2/(n-1)}} = \frac{Z_1 - \sqrt{n}\mu}{\sqrt{\sum_{i=2}^{n} Z_i{}^2/(n-1)}}$$

is a t random variable with $n - 1$ df. Since we observe the X_i's (i.e., in the $\boldsymbol{\delta}_i$ coordinate system), let us see what this quantity t looks like in terms of the X_i's. Now

$$Z_1 = \mathbf{b}_1'\mathbf{X} = \frac{\mathbf{1}'}{\|\mathbf{1}\|}\mathbf{X} = \frac{\sum_{i=1}^n X_i}{\sqrt{n}} = \sqrt{n}\bar{X}.$$

Since $\mathbf{Z}$ and $\mathbf{X}$ represent the coordinates of the same data point but with respect to different coordinate systems, the lengths of each vector are equal (Problem 2.21). Hence the sum of squares of the coordinates are equal. Now in our case,

$$\sum_{i=1}^n X_i^2 = \sum_{i=1}^n Z_i^2 = Z_1^2 + \sum_{i=2}^n Z_i^2 = (\sqrt{n}\bar{X})^2 + \sum_{i=2}^n Z_i^2$$

or

$$\sum_{i=2}^n Z_i^2 = \sum_{i=1}^n X_i^2 - n\bar{X}^2 = \sum_{i=1}^n (X_i - \bar{X})^2.$$

Finally,

$$t = \frac{\sqrt{n}(\bar{X} - \mu)}{\sqrt{\sum_{i=1}^n (X_i - \bar{X})^2/(n-1)}}.$$

SUMMARY

The Geometric Approach

(1) The random sample $X_1, \ldots, X_n$ is considered as a random vector

$$\mathbf{X} = \begin{pmatrix} X_1 \\ X_2 \\ \vdots \\ X_n \end{pmatrix}$$

taking values in n-dimensional Euclidean space with the usual inner product.

(2) Vectors are expressed using a coordinate representation; that is, each vector is written as the ordered n-tuple of its coordinates relative to a given ordered basis.

(3) Our random sample is naturally recorded and expressed with respect to the $\boldsymbol{\delta}_1, \ldots, \boldsymbol{\delta}_n$ orthonormal coordinate system.

(4) The "natural" coordinate system of the observations is not the best coordinate system for understanding the distribution theory of a random vector, and so we transform to an orthonormal coordinate system based on the geometry of the problem at hand.

(5) For ease of application we express the answer in terms of the original X_i's.

In subsequent chapters we will develop short cuts to find the final result without explicitly going through all the steps. Nevertheless, to understand why our methods work, the underlying geometry must be kept in mind.

Useful Orthonormal Coordinate Systems or (Equivalently)
Useful Orthogonal Linear Transformations

(1) If $\mathbf{X} \sim N(\boldsymbol{\mu}, \mathbf{C})$, then by the Principal Axis theorem there exists an orthogonal $\mathbf{P}$ such that $\mathbf{PCP}' = \mathbf{D}$, where $\mathbf{D}$ is diagonal and

(i) $\mathbf{PX} \sim N(\mathbf{P}\boldsymbol{\mu}, \mathbf{D})$ (Principal Axis transformation $\mathbf{P}$),
(ii) there exists an orthonormal coordinate system such that if the multivariate normal density function exists, it factors into the product of n terms, each involving only one coordinate.

(2) If $\mathbf{X} \sim N(\boldsymbol{\mu}, \sigma^2\mathbf{I})$, let $\mathbf{b}_1 = \boldsymbol{\mu}/\|\boldsymbol{\mu}\|$, $\mathbf{b}_2, \ldots, \mathbf{b}_n$ be an orthonormal basis for R^n and

$$\mathbf{P} = \begin{pmatrix} \mathbf{b}_1' \\ \mathbf{b}_2' \\ \vdots \\ \mathbf{b}_n' \end{pmatrix}$$

be the orthogonal matrix that transforms coordinate systems; then

$$\mathbf{PX} \sim N\left(\begin{pmatrix} \|\boldsymbol{\mu}\| \\ 0 \\ \vdots \\ 0 \end{pmatrix}, \sigma^2\mathbf{I}\right).$$

PROBLEMS[3]

2.1 If $\mathbf{X}^{n \times 1}$ has a mean vector $\boldsymbol{\mu}$ and covariance matrix $\mathbf{C}$, then $\mathbf{a}'\mathbf{X}$ has mean $\mathbf{a}'\boldsymbol{\mu}$ and variance $\mathbf{a}'\mathbf{Ca}$.

2.2 Show that a $N(\mu, \sigma^2)$ random variable has moment generating function $M(t) = \exp(\mu t + \frac{1}{2}t^2\sigma^2)$. [*Hint*: Complete the square.]

[3] In problems, * = difficult, † = very difficult, § = geniuses only.

2.3 From Problems 2.1, 2.2 and Definition 2.2, show that if $\mathbf{X}^{n\times 1}$ is a multinormal random variable, its moment generating function is given by

$$M(\mathbf{t}^{n\times 1}) \equiv E(e^{\mathbf{t}'\mathbf{X}}) = \exp(\boldsymbol{\mu}'\mathbf{t} + \tfrac{1}{2}\mathbf{t}'\mathbf{C}\mathbf{t}).$$

2.4§ Show that if a random vector $\mathbf{X}^{n\times 1}$ has a moment generating function $M(\mathbf{t}) = E(e^{\mathbf{t}'\mathbf{X}})$ which exists for all $\mathbf{t}$ in a neighborhood of $\mathbf{0}$, then the distribution of $\mathbf{X}$ is (uniquely) determined by the moment generating function.

2.5 Let $\mathbf{X}$ be a random vector whose moment generating function $M(\mathbf{t})$ exists in a neighborhood of $\mathbf{0}$. Show that for all nonnegative integers $m_1, \ldots, m_n$ the moment

$$E(X_1^{m_1}X_2^{m_2}\cdots X_n^{m_n}) = \frac{\partial^{m_1+\cdots+m_n}}{\partial^{m_1}t_1\cdots\partial^{m_n}t_n}(M(\mathbf{t}))\big|_{\mathbf{t}=\mathbf{0}} < \infty,$$

where $\mathbf{X}' = (X_1, \ldots, X_n)$, $\mathbf{t}' = (t_1, \ldots, t_n)$. Assume that the integration and differentiation may be interchanged.

2.6 If a symmetric matrix $\mathbf{C}$ is positive definite, show that $\mathbf{C}^{-1}$, the inverse of $\mathbf{C}$, is also positive definite. (Why does $\mathbf{C}^{-1}$ exist?)

2.7 Linear transformations of multinormal distributions are multinormal; that is, if $\mathbf{X}^{n\times 1}$ is $N(\boldsymbol{\mu}, \mathbf{C})$ and $\mathbf{T}^{m\times n}$ is a matrix, show that $\mathbf{Z} = \mathbf{TX}$ is $N(\mathbf{T}\boldsymbol{\mu}, \mathbf{TCT}')$. [*Hint*: Find the moment generating function of $\mathbf{Z}$ from the moment generating function of $\mathbf{X}$.]

2.8 (a) If $\mathbf{b}_1, \ldots, \mathbf{b}_n$ is an orthonormal basis for R^n, then any $\mathbf{X}$ is uniquely expressed in terms of this basis as

$$\mathbf{X} = X_1\mathbf{b}_1 + \cdots + X_n\mathbf{b}_n,$$

where $X_i = \mathbf{X}'\mathbf{b}_i$. (Show this.)

(b) (*Pythagorean theorem, Bessel's equality*) Show

$$\|\mathbf{X}\|^2 = \sum_{i=1}^{n} X_i^2.$$

2.9 In R^2 let two orthonormal bases $\boldsymbol{\delta}_1, \boldsymbol{\delta}_2$ and $\mathbf{b}_1, \mathbf{b}_2$ be related as shown in Figure 2.1. Find the matrix P relating the bases in terms of the angle θ.

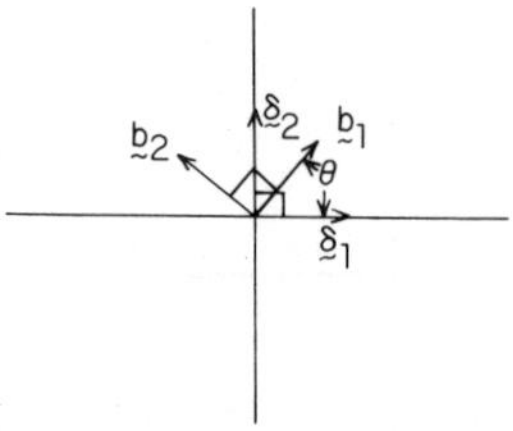

FIGURE 2.1

2.10 If $\mathbf{C}^{n \times n}$ is a real, symmetric matrix and P an orthogonal matrix such that $\mathbf{PCP}' = \mathbf{D}$, where $\mathbf{D}$ is diagonal, show that the diagonal elements of $\mathbf{D}$ are the eigenvalues of $\mathbf{C}$.

2.11 Find an orthogonal matrix $\mathbf{P}^{2 \times 2}$ such that $\mathbf{P}\left(\begin{smallmatrix} 3 & 1 \\ 1 & 3 \end{smallmatrix}\right)\mathbf{P}'$ is diagonal.

2.12 Show that a real symmetric $\mathbf{C}$ is positive (nonnegative) definite if and only if all the eigenvalues are positive (nonnegative).

2.13 (a) Show that a covariance matrix $\mathbf{C}$ is nonnegative definite.

(b) Show that a covariance matrix is positive definite if, and only if, it is nonsingular.

2.14* (*Singular normal distributions*) (a) If Z is a random variable with zero variance, show that there exists a number c such that $P(Z = c) = 1$.

(b) Let $\mathbf{X}^{n \times 1}$ be a random vector with covariance matrix of rank p, where $p < n$. Show that there is a p-dimensional subspace, say V_p, of R^n such that

$$P(\mathbf{X} - E(\mathbf{X}) \text{ in } V_p) = 1.$$

[*Hint*: Find one appropriate basis for representing $\mathbf{X}$.]

(c) Let $X_{i_1}, \ldots, X_{i_p}$ be p components from $\mathbf{X}' = (X_1, \ldots, X_n)$ whose covariance matrix has rank p. Show that $(X_{i_1}, \ldots, X_{i_p})$ determines $\mathbf{X}$. If $\mathbf{X}$ is multinormal, then $\mathbf{X}_p' = (X_{i_1}, \ldots, X_{i_p})$ has a density function (see Theorem 2.1).

2.15* Let $\mathbf{X} \sim N(\mu\mathbf{v}, \sigma^2\mathbf{I})$. Show that the distribution of $\mathbf{X}$ given $\mathbf{X}'\mathbf{v}$ does not depend on μ. Thus, $\mathbf{X}'\mathbf{v}$ contains all the information about μ in the situation of the text (page 13).

2.16 (a) Let $\mathbf{X}' = (X_1, X_2)$ have a bivariate normal distribution. Show that X_1 and X_2 are independent if and only if $\operatorname{cov}(X_1, X_2)$ is zero.

(b) Let $\mathbf{X}' = (\mathbf{X}_1', \mathbf{X}_2')$ be multinormal $\mathbf{X}_1' = (X_1, \ldots, X_k)$, $\mathbf{X}_2' = (X_{k+1}, \ldots, X_n)$. Show that $\mathbf{X}_1$ and $\mathbf{X}_2$ are independent vectors if the "covariance matrix" $\mathbf{C}_{12}^{k \times (n-k)}$ has all zero entries where $\mathbf{C}_{12}$ has the covariance between the components of $\mathbf{X}_1$ and those of $\mathbf{X}_2$ (covariance matrix for $\mathbf{X}$,

$$\mathbf{C} = \begin{pmatrix} \mathbf{C}_{11} & \mathbf{C}_{12} \\ \mathbf{C}_{12}' & \mathbf{C}_{22} \end{pmatrix}).$$

2.17 Let $\mathbf{X}^{n \times 1} \sim N(\boldsymbol{\mu}, \mathbf{C})$ and $\mathbf{a}, \mathbf{b}$ be vectors in R^n. Show that $\operatorname{cov}(\mathbf{a}'\mathbf{X}, \mathbf{b}'\mathbf{X}) = \mathbf{a}'\mathbf{C}\mathbf{b}$.

2.18 (a) Describe the Gram–Schmidt orthogonalization process.

(b) Let $\mathbf{v}_1, \ldots, \mathbf{v}_k$ be nonzero orthogonal vectors of R^n ($k < n$). Show that there is an orthonormal basis $\mathbf{b}_1, \ldots, \mathbf{b}_n$ such that

$$\mathbf{b}_i = \mathbf{v}_i / \|\mathbf{v}_i\|, \qquad i = 1, 2, \ldots, k.$$

2.19 Let $\{\boldsymbol{\delta}_1, \ldots, \boldsymbol{\delta}_n\}$ and $\{\mathbf{b}_1, \ldots, \mathbf{b}_n\}$ be two orthonormal coordinate systems (expressed as n-tuples with respect to the $\boldsymbol{\delta}_i$ coordinate system). Let $\mathbf{X}$ and $\mathbf{Z}$ be expressions of a vector with respect to the $\boldsymbol{\delta}_i$ and $\mathbf{b}_i$ coordinate systems respectively, and $\mathbf{Z} = \mathbf{PX}$. Show that $\mathbf{P}$ expressed in terms of its n rows is

$$\mathbf{P} = \begin{pmatrix} \mathbf{b}_1' \\ \vdots \\ \mathbf{b}_n' \end{pmatrix}.$$

2.20 In R^2, consider two coordinate systems $\boldsymbol{\delta}_1, \boldsymbol{\delta}_2$ and $\mathbf{b}_1, \mathbf{b}_2$ as shown in Figure 2.2. Let $\mathbf{X}$ and $\mathbf{Y}$ be as shown in the figure.

(a) Find the representations of $\mathbf{X}$ and $\mathbf{Y}$ in the $\{\boldsymbol{\delta}_1, \boldsymbol{\delta}_2\}$ coordinate system. Compute $\mathbf{X}'\mathbf{Y}$ from this representation.

(b) Find the representations of $\mathbf{X}$ and $\mathbf{Y}$ in the $\{\mathbf{b}_1, \mathbf{b}_2\}$ coordinate system. Compute $\mathbf{X}'\mathbf{Y}$ from this representation.

(c) Find the matrix $\mathbf{P}$ that relates representations in the two coordinate systems.

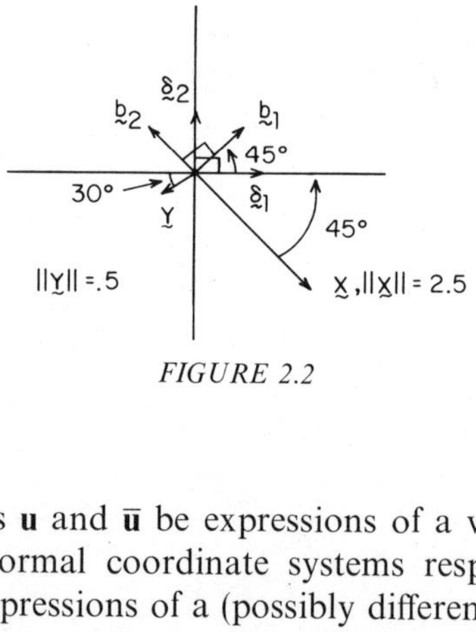

FIGURE 2.2

2.21 Let the n-tuples $\mathbf{u}$ and $\bar{\mathbf{u}}$ be expressions of a vector with respect to the $\boldsymbol{\delta}_i$ and $\mathbf{b}_i$ orthonormal coordinate systems respectively, and let the n-tuples $\mathbf{v}$ and $\bar{\mathbf{v}}$ be expressions of a (possibly different) vector with respect to the $\boldsymbol{\delta}_i$ and $\mathbf{b}_i$ coordinate systems respectively; then show that inner products are the same regardless of which system is used, i.e., $\mathbf{u}'\mathbf{v} = \bar{\mathbf{u}}'\bar{\mathbf{v}}$. Hence the length of a vector is the same in either coordinate system (of course).

2.22 Data for Example 2.1 are given in Table 2.2.

(a) Compute the t-value for group I, nipple rolling. How strong is the evidence for a difference?

(b) Same as for question (a), but for groups II and III.

TABLE 2.2

Subjective Difference between Treated and Untreated Breasts of Three Groups of 19 Subjects Each as Measured by t-Test

	Group		
Subject	I Nipple rolling	II Masse cream	III Expression of colostrum
1	−.525	.026	−.006
2	.172	.739	.000
3	−.577	−.095	−.257
4	.200	−.040	−.070
5	.040	.006	.107
6	−.143	−.600	.362
7	.043	.007	−.263
8	.010	.008	.010
9	.000	.000	−.080
10	−.522	−.100	−.010
11	.007	.000	.048
12	−.122	.000	.300
13	−.040	.060	.182
14	.000	−.180	−.378
15	−.100	.000	−.075
16	.050	.040	−.040
17	−.575	.080	−.080
18	.031	−.450	−.100
19	−.060	.000	−.020

2.23 Consider the data of Example 2.3. Compute the paired t-test value comparing the 24-hr before-treatment value with the 8- (48) hr values. Is the mean statistically significantly nonzero at a significance level of .10, .05, .01? Do you think the variances of the average apneic episodes per hour are the same for the 24-hr control period and the 8-hr treatment periods? What is your reason?

2.24 [Abstract from J. C. Dobson, E. Kushida, M. Williamson, and E. Friedman, Phenylketonuria, metabolic disorder, mental retardation, inborn errors of amino acid metabolism, *Pediatrics* **58**, 53–58 (1976).] "Thirty-six patients with a confirmed diagnosis of phenylketonuria (PKU) were identified and placed on dietary therapy before reaching 121 days of age. These children are currently between 4 and 6 years old, and have been given the Stanford–Binet Intelligence Scale. Subsequently, their normal siblings of closest age were selected as matched-pair controls and were also tested with the Stanford–Binet. The mean age of the PKU sample when tested was 50.0 months old, and 50.9 months for the normal controls. The 36 index patients scored a mean IQ of 94 and their nonaffected siblings obtained a mean IQ of 99. This five-point mean difference was significant at the .02 level, and suggests the presence of a minimal intellectual impairment associated with PKU, even when treatment begins early and is rigorously monitored."

Compute the paired t value for comparing the PKU cases and their siblings. (See Table 2.3.) What can you say about the p value?

TABLE 2.3

Matched pairs code	Stanford–Binet IQ PKU index case	1972 Norms sibling	Difference between IQs
1	89	77	PKU + 12
2	98	110	PKU − 12
3	116	94	PKU + 22
4	67	91	PKU − 24
5	128	122	PKU + 6
6	81	94	PKU − 13
7	96	121	PKU − 25
8	116	114	PKU + 2
9	110	88	PKU + 22
10	90	91	PKU − 1
11	76	99	PKU − 23
12	71	93	PKU − 22
13	100	104	PKU − 4
14	108	102	PKU + 6
15	74	82	PKU − 8
16	94	115	PKU − 21
17	82	90	PKU − 8
18	91	105	PKU − 14
19	102	110	PKU − 8
20	74	107	PKU − 33
21	86	78	PKU + 8
22	98	93	PKU + 5
23	99	102	PKU − 3
24	82	94	PKU − 12
25	103	105	PKU − 2
26	96	103	PKU − 7
27	63	77	PKU − 14
28	72	86	PKU − 14
29	106	120	PKU − 14
30	127	129	PKU − 2
31	88	87	PKU + 1
32	73	78	PKU − 5
33	79	94	PKU − 15
34	88	68	PKU + 20
35	126	123	PKU + 3
36	119	111	PKU + 8

3 TWO-SAMPLE t-TEST

Consider now the same type of data as that in Chapter 2, except that samples are taken from two distinct populations; that is, let $X_1, \ldots, X_n$ be a random sample of $N(\mu, \sigma^2)$ random variables and let (independently) $Y_1, \ldots, Y_m$ be a random sample of $N(\omega, \sigma^2)$ random variables. Suppose μ, ω, and σ^2 are unknown. Note that although the means may differ, the variance σ^2 is assumed to be the same in both populations. In this situation, tests comparing μ and ω and confidence intervals for $\mu - \omega$ are computed using the fact that

$$t = \frac{\bar{X} - \bar{Y} - (\mu - \omega)}{\sqrt{\dfrac{(n-1)s_x^2 + (m-1)s_y^2}{n+m-2}} \sqrt{\dfrac{1}{n} + \dfrac{1}{m}}} \qquad (1)$$

is a t random variable with $n + m - 2$ df, where

$$\bar{X} = \frac{1}{n}\sum_{i=1}^{n} X_i, \qquad \bar{Y} = \frac{1}{m}\sum_{i=1}^{m} Y_i,$$

$$s_x^2 = \frac{1}{n-1}\sum_{i=1}^{n}(X_i - \bar{X})^2, \qquad s_y^2 = \frac{1}{m-1}\sum_{i=1}^{m}(Y_i - \bar{Y})^2.$$

Using the results of Chapter 2 and the fact that a sum of independent χ^2 random variables is again a χ^2 random variable (Problem 3.1), it is easy to show that (1) does indeed define a t random variable with $n + m - 2$ df. Before continuing with our theoretical work, consider some examples using the two-sample t-distribution.

Example 3.1 [M. Cadie, F. J. Nye, and P. Storey, Anxiety and depression after infectious mononucleosis, *British Journal of Psychiatry* **128**, 559–561

(1976)] There is a widespread clinical impression that infectious mononucleosis (IM) is often followed by prolonged states of anxiety and/or depression. Since little has been published on the subject, Cadie *et al.* investigated this clinical impression.

Thirty-six patients who had had IM were followed for one year and then filled out the Middlesex Hospital Questionnaire (MHQ). The MHQ is a brief self-rating inventory intended to cover the full clinical range of neurotic illness. It provides scores on six scales: anxiety, phobic, obsessional, somatic, depression, and hysteria. These scales can be taken as reflecting the patient's clinical state in symptomatic terms.

Of the 36 patients completing the MHQ, 20 were females and 16 males. The mean age of the women was 24 years (SD = 7) and of the men 23 years (SD = 5). As there were no exact controls available, control samples were taken from a General Practice population aged 20 to 24. Table 3.1 shows the mean scores, the standard deviations, and number of observations for the six scales for men and women and control and IM patients separately.

TABLE 3.1

Middlesex Hospital Questionnaire Results after Infectious Mononucleosis and in Controls.[a] *Mean Scores (and Standard Deviations)*

	Men			Women		
	IM patients ($N = 16$) Mean	Controls ($N = 32$) Mean	t	IM patients ($N = 20$) Mean	Controls ($N = 26$) Mean	t
Anxiety	3.5 (3.1)	3.1 (2.9)	.47	6.3 (3.6)	3.5 (3.1)	2.73[b]
Phobic	3.3 (2.2)	2.7 (1.9)	.95	4.9 (3.1)	3.9 (1.9)	1.28
Obsessional	6.4 (4.2)	4.7 (2.6)	1.49	4.9 (2.1)	5.2 (2.3)	.40
Somatic	3.6 (2.1)	3.3 (2.8)	.34	4.0 (2.2)	2.3 (1.7)	2.79[b]
Depression	2.2 (2.4)	2.2 (2.5)	.00	3.7 (2.9)	1.9 (1.6)	2.49[b]
Hysteria	4.9 (3.5)	4.8 (3.7)	.10	5.9 (3.2)	5.3 (4.1)	.55

[a] The control samples were taken from the General Practice population aged 20 to 24; significance levels are based on a two-tailed test.

[b] $p < .02$.

Two sample *t*-tests were calculated to compare the IM patients and controls by sex. The authors conclude that there are only trivial differences between the male patients and controls, but the female patients score significantly higher on anxiety, somatic symptoms, and depression. This study, although small, does support the view that IM is significantly often followed by depression, although only in women.

Example 3.2 [B. Lester, Spectrum analysis of the cry sounds of the well-nourished and malnourished infants, *Child Development* **47**, 237–241 (1976)][1] Both behavioral and acoustic measures have been used to differentiate the cry of the normal from that of the abnormal infant and many investigators have pointed to the diagnostic utility of the cry as a sensitive indicator of the integrity of the central nervous system (CNS). Since the cry originates in the CNS, it is reasonable to suggest that substantial deviations from the normal cry pattern may be indicative of CNS dysfunction. Lester compared behavioral and acoustic parameters of the cry sounds between normal infants and infants possibly suffering from CNS insult due to malnutrition.

The subjects were 24 Guatemalan male infants who were full term, full birth weight, clinically normal at birth, and suffered no major illnesses as of the time of testing. Twelve infants were well-nourished and 12 infants were malnourished. Table 3.2 gives various descriptive characteristics of the sample for each group of infants. The malnourished infants were significantly ($p < .001$) lighter, shorter, had a smaller head circumference, and were lighter for their age than the well-nourished infants. There was no significant differences in their ages.

TABLE 3.2

Descriptive Characteristics of Sample

	Group			
	Well nourished ($N = 12$)		Malnourished ($N = 12$)	
Variable	$\bar{X}$	SD	$\bar{X}$	SD
Age (months)	12.37	1.44	12.20	1.26
Weight (kilos)	9.26	.45	6.69	.79
Percentage of weight for age	84.92	3.37	62.69	7.00
Height (cm)	71.78	2.92	66.23	3.43
Head circumference	45.17	.98	42.94	1.50

The cry sounds were analyzed with a real time spectrum analyzer. The analysis was performed by the author without knowledge, that is, "blind," of the nutritional status of the infant. Five measures, two behavioral and three acoustic, were made from each cry segment. The behavioral measures were the duration of the initial cry and the latency to the onset of the next cry sound. The acoustic measures were the fundamental frequency of the initial cry sound, the amplitude or intensity of the fundamental frequency,

TABLE 3.3

Means and Standard Deviations of Cry Measures in Well-Nourished and Malnourished Infants

	Group			
	Well nourished ($N = 12$)		Malnourished ($N = 12$)	
Cry measure	$\bar{X}$	SD	$\bar{X}$	SD
Duration of cry (sec)	1.52	.39	2.66	1.15
Latency to next cry (sec)	.47	.26	1.80	.69
Frequency of fundamental (Hz)	308.00	32.18	479.77	25.61
Amplitude of fundamental (dB)	50.31	7.00	38.16	5.47
Number of peaks	4.5	1.20	2.25	.75

and the number of peaks or major shifts in the frequency of the initial cry. The mean and standard deviation for each cry measure for the well-nourished and malnourished infants are presented in Table 3.3.

There were significant differences between the groups for all cry variables. In sum, the cry of the malnourished infant can be characterized by an initial long sound that is high in pitch, low in intensity, and with a long interval to the next cry sound. The malnourished cry also appeared arrhythmical as indicated by the small number of peaks in the initial cry sound. The results from this study show a rather striking similarity between the cry of malnourished infants and the cries of infants suffering from CNS dysfunction reported in other studies. If aberrations of the cry pattern are indicative of CNS dysfunction, these findings may suggest that the regulatory function of the CNS is affected by nutritional insult.

Before proceeding further with the two-sample t-test, we digress on the topic of estimation of parameters. For $X_1, \ldots, X_n$, a random sample from a distribution with mean μ and variance σ^2, it is well known that the sample mean $\bar{X}$ is an unbiased estimator of μ and the sample variance $s_x^{\,2}$ is an unbiased estimator of σ^2. In our derivation of the one-sample t-test, we implicitly showed that for a random sample from a normal distribution, the sample mean and sample variance are independent. We shall show how to use the geometry of the problem at hand to obtain independent estimators of various parameters.

Consider our previous problem $\mathbf{X} \sim N(\mu\mathbf{1}, \sigma^2\mathbf{I})$, where the expected value of the observation vector lies in the one-dimensional subspace spanned by the $\mathbf{1}$ vector, that is, in the $\mathbf{1}$ direction. One intuitive estimate of the parameter μ is that value of the parameter which results in a mean vector nearest the observation vector. To find the mean vector nearest the observation vector, we need the concept of the orthogonal projection of a vector onto a given subspace and a method for finding this vector.

The orthogonal projection of a vector **X** onto a subspace S is given geometrically by dropping a perpendicular from the vector onto the subspace and is that vector in S which is nearest to **X** (see Figure 3.1). Note that **X** minus the orthogonal projection of **X** is perpendicular to S. We shall use this property in defining the orthogonal projection operator and the orthogonal projection of a vector onto a subspace.

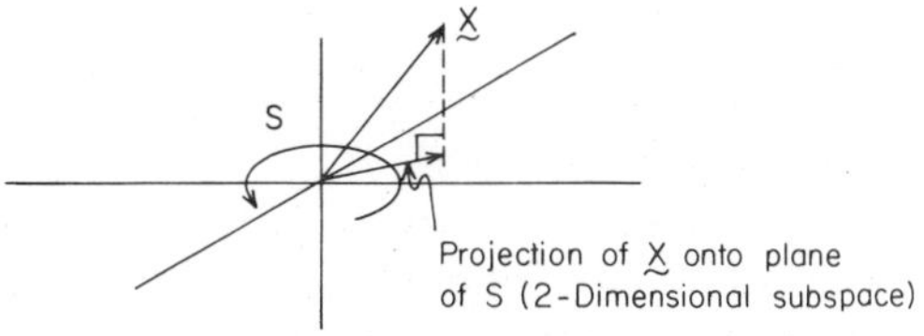

FIGURE 3.1

DEFINITION 3.1 The *orthogonal projection operator* onto S, a subspace of a vector space V, denoted by $\mathbf{P}_S$, assigns to each vector **X** in V the vector $\mathbf{P}_S\mathbf{X}$ in S, called the *orthogonal projection of* **X** onto S, such that $\mathbf{X} - \mathbf{P}_S\mathbf{X}$ is orthogonal to S.

Since we shall be dealing only with orthogonal projection operators, we will often drop the adjective orthogonal and even the term operator and just speak of projections.

Problems 3.2–3.6 along with Appendix 1 give some needed background facts about projections. Problem 3.4 basically says that the projection operator can be written as an explicit function of an orthonormal basis for the subspace we are projecting onto; that is, if $\mathbf{b}_1, \ldots, \mathbf{b}_k$ is an orthonormal basis for a subspace S, then $\mathbf{P}_S = \sum_{i=1}^{k} \mathbf{b}_i\mathbf{b}_i'$. The *outer product* of a vector **b** is defined as the matrix $\mathbf{bb}'$, so that the projection operator onto a subspace is given by the sum of the outer products of the orthonormal basis vectors for that subspace.

For $\mathbf{X} \sim N(\mu\mathbf{1}, \sigma^2\mathbf{I})$, the expected value of the observation vector lies in the one-dimensional subspace spanned by the **1** vector. Dividing the **1** vector by its length $\|\mathbf{1}\|$ gives us an orthonormal basis for this subspace, so that the projection operator onto this subspace is $(\mathbf{1}/\|\mathbf{1}\|)(\mathbf{1}'/\|\mathbf{1}\|)$. The projection of the observation vector **X** onto this subspace is

$$\frac{\mathbf{1}}{\|\mathbf{1}\|}\frac{\mathbf{1}'\mathbf{X}}{\|\mathbf{1}\|} = \mathbf{1}\frac{\mathbf{1}'\mathbf{X}}{\|\mathbf{1}\|^2} = \mathbf{1}\frac{\sum_{i=1}^{n} X_i}{n} = \mathbf{1}\bar{X}.$$

Our estimate of μ, denoted $\hat{\mu}$, is that value of the parameter which results in a mean vector nearest the observation vector; that is, we solve $\hat{\mu}\mathbf{1} = \bar{X}\mathbf{1}$ for $\hat{\mu}$ so that obviously $\hat{\mu} = \bar{X}$.

The problem with using $\bar{X}$ to estimate μ or to test hypotheses about μ is that the accuracy of the estimate depends on the unknown quantity σ^2 [remember $\bar{X} \sim N(\mu, \sigma^2/n)$ for a normal random sample]. Since only the **1** direction of R^n was used to estimate μ, there should be "directions left" for estimating σ^2. We used the orthonormal coordinate system $\mathbf{b}_1 = \mathbf{1}/\|\mathbf{1}\|$, $\mathbf{b}_2, \ldots, \mathbf{b}_n$ to derive our one-sample t-test. The $n-1$ orthonormal vectors $\mathbf{b}_2, \ldots, \mathbf{b}_n$ span a $(n-1)$-dimensional subspace of R^n consisting of all vectors orthogonal to the **1** direction; that is, the orthogonal complement of the subspace spanned by the **1** vector and certainly offer us "directions left" for estimating σ^2.

If S denotes the one-dimensional subspace spanned by the **1** vector and $S^\perp$ its orthogonal complement, then $R^n = S \oplus S^\perp$, where $\oplus$ denotes the direct sum of the subspaces S and $S^\perp$. Appendix 1 reviews the concept of the direct sum of subspaces and its relation to orthogonal projections.

We used the orthogonal projection of the data vector onto S to estimate μ so we might try projecting the data vector onto $S^\perp$ in our search for an estimate of σ^2. Theorem *A*26 says that $\mathbf{I} - \mathbf{P}_S$ is the orthogonal projection onto $S^\perp$ if $\mathbf{P}_S$ is the projection onto S. Therefore $(\mathbf{I} - \mathbf{P}_S)\mathbf{X} = \mathbf{X} - \mathbf{P}_S\mathbf{X} = \mathbf{X} - \bar{X}\mathbf{1}$ is the projection of the data vector onto $S^\perp$. (This is geometrically obvious from Figure 3.1.) Now

$$\mathbf{X} - \bar{X}\mathbf{1} = \begin{pmatrix} X_1 - \bar{X} \\ X_2 - \bar{X} \\ \vdots \\ X_n - \bar{X} \end{pmatrix},$$

and we notice that the squared length of this vector is $\sum_{i=1}^{n} (X_i - \bar{X})^2$ which when divided by $n-1$, the dimensionality of the subspace we are projecting onto, is an unbiased estimate of σ^2; that is,

$$E\left(\frac{\|\mathbf{P}_{S^\perp}\mathbf{X}\|^2}{\dim S^\perp}\right) = \sigma^2,$$

where $\dim S^\perp$ is the dimension of $S^\perp$.

Note that our estimator of μ, $\bar{X}$, can be considered a function of $\bar{X}\mathbf{1}$, the projection of the data vector onto S, and that our estimator of σ^2, $s^2 = \|\mathbf{P}_{S^\perp}\mathbf{X}\|^2/\dim S^\perp$, is a function of the projection of the data vector onto $S^\perp$. The key geometric idea in the analysis of variance will be that of projecting the data vector $\mathbf{X}$ onto orthogonal subspaces.

DEFINITION 3.2 Subspaces $S_1, S_2, \ldots, S_k$ are orthogonal if any two vectors chosen from two distinct subspaces are orthogonal.

Since $\bar{X}$ and s^2 are independent for the normal random sample and functions of independent random vectors are independent, one might wonder if the random vectors $\mathbf{P}_S\mathbf{X}$ and $\mathbf{P}_{S^\perp}\mathbf{X}$ are independent. *This is another key idea: for multinormal observations with a covariance matrix of the form $\sigma^2\mathbf{I}$, orthogonality and independence are equivalent.* This idea is made precise here.

THEOREM 3.1 Let $S_1, S_2, \ldots, S_k$ be orthogonal subspaces of R^n with $\mathbf{X} \sim N(\boldsymbol{\mu}, \sigma^2\mathbf{I})$ taking values in R^n. (Note that $\sigma^2\mathbf{I}$ is the normal covariance matrix.) Let $\mathbf{P}_i$ be the (orthogonal) projection onto S_i, $i = 1, \ldots, k$. Then $\mathbf{P}_i\mathbf{X}$, $i = 1, \ldots, k$, are *independent* random vectors and $\mathbf{P}_i\mathbf{X} \sim N(\mathbf{P}_i\boldsymbol{\mu}, \sigma^2\mathbf{P}_i)$.

Proof Geometrically, we would like a direct sum decomposition of R^n by orthogonal subspaces, since this particular kind of direct sum decomposition is equivalent to the study of orthogonal projections. Theorem A19 says that an orthogonal sum of subspaces is a direct sum, but we have not assumed that $S_1 \oplus S_2 \oplus \cdots \oplus S_k = R^n$. $S_1 \oplus S_2 \oplus \cdots \oplus S_k$ may be a proper subspace of R^n. If so, let S_{k+1} be the orthogonal complement of $S_1 \oplus S_2 \oplus \cdots \oplus S_k$ so that $S_1 \oplus S_2 \oplus \cdots \oplus S_k \oplus S_{k+1} = R^n$, let $\mathbf{P}_{k+1}$ be the projection onto S_{k+1} and let $k' = k + 1$. Otherwise let $k' = k$. Let $d_i = \dim S_i$ and choose $\mathbf{b}_1{}^i, \ldots, \mathbf{b}_{d_i}^i$ an orthonormal basis for S_i for $i = 1, \ldots, k'$. Now the set of all the $\mathbf{b}_j{}^i$'s, $j = 1, \ldots, d_i$, $i = 1, \ldots, k'$, is an orthonormal basis for R^n. For all $\mathbf{X}$ in R^n,

$$\mathbf{X} = \sum_{i=1}^{k'} \sum_{j=1}^{d_i} (\mathbf{X}'\mathbf{b}_j{}^i)\mathbf{b}_j{}^i.$$

But this says (Problem 3.4)

$$\mathbf{X} = \mathbf{P}_1\mathbf{X} + \cdots + \mathbf{P}_{k'}\mathbf{X}.$$

Geometrically, our observation vector $\mathbf{X}$ has been expressed as the sum of orthogonal vectors. (Remember the Pythagorean theorem; this is the higher-dimensional analogue of the geometry involved in the Pythagorean theorem for triangles with a right angle.)

To show independence of the distributions, we use moment generating functions (Problem 3.8). The joint moment generating function of $\mathbf{P}_1\mathbf{X}, \ldots, \mathbf{P}_{k'}\mathbf{X}$ is

$$\begin{aligned} M(\mathbf{t}_1, \ldots, \mathbf{t}_{k'}) &= E\left(\exp\left[\sum_{i=1}^{k'} \mathbf{t}_i'\mathbf{P}_i\mathbf{X}\right]\right) = E\left(\exp\left[\left(\sum_{i=1}^{k'} \mathbf{t}_i'\mathbf{P}_i\right)\mathbf{X}\right]\right) \\ &= \exp\left[\left(\sum_{i=1}^{k'} \mathbf{t}_i'\mathbf{P}_i\right)\boldsymbol{\mu} + \tfrac{1}{2}\left(\sum_{i=1}^{k'} \mathbf{t}_i'\mathbf{P}_i\right)\sigma^2\mathbf{I}\left(\sum_{i=1}^{k'} \mathbf{t}_i'\mathbf{P}_i\right)'\right]. \end{aligned}$$

Using $\mathbf{P}_i\mathbf{P}_i' = \mathbf{P}_i$ and $\mathbf{P}_i\mathbf{P}_j' = \mathbf{0}$ for $i \neq j$ (Problem 3.6), we obtain

$$M(\mathbf{t}_1, \ldots, \mathbf{t}_{k'}) = \exp\left[\sum_{i=1}^{k'} (\mathbf{t}_i'\mathbf{P}_i\boldsymbol{\mu} + \tfrac{1}{2}\sigma^2\mathbf{t}_i'\mathbf{P}_i\mathbf{t}_i)\right]$$

$$= \prod_{i=1}^{k'} \exp[\mathbf{t}_i'\mathbf{P}_i\boldsymbol{\mu} + \tfrac{1}{2}\mathbf{t}_i'\sigma^2\mathbf{P}_i\mathbf{t}_i].$$

Recognizing the terms in the last product as $N(\mathbf{P}_i\boldsymbol{\mu}, \sigma^2\mathbf{P}_i)$ moment generating functions, we see that the proof is complete. □

Returning to our two-sample t-test, it is easily shown that the data vector $\mathbf{X}$ has the following multinormal distribution in R^{n+m}:

$$\mathbf{X} = \begin{pmatrix} X_1 \\ \vdots \\ X_n \\ Y_1 \\ \vdots \\ Y_m \end{pmatrix} \sim N\left(\begin{pmatrix} \mu \\ \vdots \\ \mu \\ \omega \\ \vdots \\ \omega \end{pmatrix} \begin{matrix} \left.\vphantom{\begin{matrix}\mu\\ \vdots\\ \mu\end{matrix}}\right\} n \text{ positions} \\ \left.\vphantom{\begin{matrix}\omega\\ \vdots\\ \omega\end{matrix}}\right\} m \text{ positions}\end{matrix}, \sigma^2\mathbf{I}\right).$$

Note that the mean vector is equal to $\mu\mathbf{v}_1 + \omega\mathbf{v}_2$, where $\mathbf{v}_1$ has n ones and then m zeros as entries and $\mathbf{v}_2$ has n zeros and then m ones as entries. Also note that $\mathbf{v}_1$ and $\mathbf{v}_2$ are orthogonal and hence linearly independent vectors.

The expected value of the observation vector lies in the subspace spanned by the basis vectors $\mathbf{v}_1$ and $\mathbf{v}_2$. Intuitive estimates of the parameters μ and ω are those values of the parameters which result in a mean vector nearest the observation vector. These estimates are unique. (Why does this follow from Problem 3.2 and independence of $\mathbf{v}_1$ and $\mathbf{v}_2$?) Now $\mathbf{b}_1 = \mathbf{v}_1/\|\mathbf{v}_1\|$ and $\mathbf{b}_2 = \mathbf{v}_2/\|\mathbf{v}_2\|$ are an orthonormal basis for the subspace in which the mean vector lies. The projection operator onto this subspace is $\mathbf{b}_1\mathbf{b}_1' + \mathbf{b}_2\mathbf{b}_2'$, and the projection of the data vector is

$$\mathbf{b}_1\mathbf{b}_1'\mathbf{X} + \mathbf{b}_2\mathbf{b}_2'\mathbf{X} = \frac{\mathbf{v}_1}{\|\mathbf{v}_1\|}\frac{\mathbf{v}_1'\mathbf{X}}{\|\mathbf{v}_1\|} + \frac{\mathbf{v}_2}{\|\mathbf{v}_2\|}\frac{\mathbf{v}_2'\mathbf{X}}{\|\mathbf{v}_2\|} = \mathbf{v}_1\frac{\mathbf{v}_1'\mathbf{X}}{\|\mathbf{v}_1\|^2} + \mathbf{v}_2\frac{\mathbf{v}_2'\mathbf{X}}{\|\mathbf{v}_2\|^2}$$

$$= \mathbf{v}_1\left(\frac{\sum_{i=1}^n X_i}{n}\right) + \mathbf{v}_2\left(\frac{\sum_{j=1}^m Y_j}{m}\right) = \mathbf{v}_1\bar{X} + \mathbf{v}_2\bar{Y}.$$

Obviously $\hat{\mu} = \bar{X}$ and $\hat{\omega} = \bar{Y}$. As $\mathbf{v}_1$ and $\mathbf{v}_2$ are orthogonal, by our previous theorem the projections of the data vector onto the "two directions" are independent random vectors. $\bar{X}$ and $\bar{Y}$ are functions of these independent random vectors, and hence $\bar{X}$ and $\bar{Y}$ are independent estimators. Since

$\bar{X} = \mathbf{v}_1'\mathbf{X}/\|\mathbf{v}_1\|^2$, also note that

$$E(\bar{X}) = E\left(\frac{\mathbf{v}_1'\mathbf{X}}{\|\mathbf{v}_1\|^2}\right) = \frac{\mathbf{v}_1'E(\mathbf{X})}{\|\mathbf{v}_1\|^2} = \frac{\mathbf{v}_1'(\mu\mathbf{v}_1 + \omega\mathbf{v}_2)}{\|\mathbf{v}_1\|^2} = \mu\frac{\mathbf{v}_1'\mathbf{v}_1}{\|\mathbf{v}_1\|^2} = \mu,$$

so that $\bar{X}$ is an unbiased estimator of μ. Similarly, $\bar{Y}$ can be shown to be an unbiased estimator of ω.

The $(n + m - 2)$-dimensional subspace orthogonal to $\mathbf{v}_1$ and $\mathbf{v}_2$ may be used to estimate σ^2. How should we use this subspace to estimate σ^2? Changing to a more convenient orthonormal coordinate system, we let

$$\mathbf{b}_1 = \mathbf{v}_1/\|\mathbf{v}_1\|, \quad \mathbf{b}_2 = \mathbf{v}_2/\|\mathbf{v}_2\|, \quad \mathbf{b}_3, \ldots, \mathbf{b}_N \qquad (N = n + m)$$

be an orthonormal basis for R^N. As before, we see that if we let $Z_i = \mathbf{b}_i'\mathbf{X}$, then

$$\mathbf{Z} = \begin{pmatrix} Z_1 \\ Z_2 \\ Z_3 \\ \vdots \\ Z_N \end{pmatrix} \sim N\left(\begin{pmatrix} \mu\sqrt{n} \\ \omega\sqrt{m} \\ 0 \\ \vdots \\ 0 \end{pmatrix}, \sigma^2\mathbf{I}\right).$$

Thus $\sum_{i=3}^N (Z_i^2/\sigma^2)$ is a χ^2 random variable with $n + m - 2$ df. A test for $\mu = \omega$ or equivalently $\mu - \omega = 0$ is now clear. A t random variable with $n + m - 2$ df is given by

$$t = \frac{[(n^{-1/2}Z_1 - m^{-1/2}Z_2) - (\mu - \omega)]/\sqrt{(\sigma^2/n) + (\sigma^2/m)}}{\sqrt{\sum_{i=3}^N (Z_i^2/\sigma^2)/(n + m - 2)}}.$$

(Can you supply the details?) (Note the σ^2 cancel.) Let us see what this quantity looks like in terms of the observations. The length of the data vector is the same in either coordinate system, so that

$$\|\mathbf{X}\|^2 = \|\mathbf{Z}\|^2 \qquad \text{or} \qquad \|\mathbf{X}\|^2 = \sum_{i=1}^n X_i^2 + \sum_{j=1}^m Y_j^2 = Z_1^2 + Z_2^2 + \sum_{i=3}^N Z_i^2,$$

and as $Z_1 = \sqrt{n}\bar{X}$ and $Z_2 = \sqrt{m}\bar{Y}$, we see that

$$\sum_{i=3}^N Z_i^2 = \sum_{i=1}^n (X_i - \bar{X})^2 + \sum_{j=1}^m (Y_j - \bar{Y})^2.$$

The obvious substitutions yield equation (1).

Since $\sum_{i=3}^N (Z_i^2/\sigma^2) \sim \chi^2_{n+m-2}$ and recalling that the mean of a χ^2 random variable equals its degrees of freedom, we have $E(\sum_{i=3}^N Z_i^2)/\sigma^2 = n + m - 2$ or $E(\sum_{i=3}^N Z_i^2)/(n + m - 2) = \sigma^2$. Note that (Problem 3.10) the projection of $\mathbf{X}$ onto the orthogonal complement of the subspace S spanned by $\mathbf{v}_1$

and $\mathbf{v}_2$ is given by $\mathbf{P}_{S^\perp}\mathbf{X} = \mathbf{b}_3(\mathbf{b}_3'\mathbf{X}) + \cdots + \mathbf{b}_N(\mathbf{b}_N'\mathbf{X})$. Further, $\sum_{i=3}^N Z_i^2 = \sum_{i=3}^N (\mathbf{b}_i'\mathbf{X})^2 = ||\mathbf{P}_{S^\perp}\mathbf{X}||^2$, so that once again an unbiased estimate of σ^2 is given by $||\mathbf{P}_{S^\perp}\mathbf{X}||^2/\dim S^\perp$. Note that this estimate is independent of $\bar{X}$ and $\bar{Y}$. (Why?)

Since $||\mathbf{P}_{S^\perp}\mathbf{X}||^2/\sigma^2$ for the preceding case turned out to have a χ^2-distribution with $\dim S^\perp$ df, let us look in general at the distribution of $||\mathbf{PX}||^2/\sigma^2$ for an arbitrary projection $\mathbf{P}$. This will turn out to be a noncentral χ^2-distribution in our applications.

LEMMA and DEFINITION 3.3 Let $\mathbf{X} \sim N(\boldsymbol{\mu}, \sigma^2\mathbf{I})$ take values in R^n. Then the distribution of $||\mathbf{X}||^2/\sigma^2$ only depends on the parameter $\delta = ||\boldsymbol{\mu}||/\sigma$ called the *noncentrality parameter.*[2] The distribution is called the *noncentral chi-squared distribution* with n df and noncentrality parameter δ, and is denoted $\chi_n{}^2(\delta)$. If $\delta = 0$, the noncentral χ^2-distribution is the usual or central χ^2-distribution, i.e., $\chi_n{}^2(0) = \chi_n{}^2$.

Proof (a) If $\delta = 0$, then $\boldsymbol{\mu} = \mathbf{0}$ and the usual χ^2-distribution is immediate as the sum of squares of n independent $N(0, 1)$ random variables.

(b) If $\delta \neq 0$, then $||\boldsymbol{\mu}|| \neq 0$; so let $\mathbf{b}_1 = \boldsymbol{\mu}/||\boldsymbol{\mu}||, \mathbf{b}_2, \ldots, \mathbf{b}_n$ be an orthonormal basis for R^n. Let $Z_i = \mathbf{b}_i'\mathbf{X}$ and $\mathbf{Z}' = (Z_1, \ldots, Z_n)$. Then the distribution of $\mathbf{Z}$ is easily shown to be

$$\mathbf{Z} \sim N\left(\begin{pmatrix} ||\boldsymbol{\mu}|| \\ 0 \\ \vdots \\ 0 \end{pmatrix}, \sigma^2\mathbf{I}\right).$$

Also,

$$\sum_{i=1}^n \frac{X_i^2}{\sigma^2} = \frac{Z_1^2}{\sigma^2} + \sum_{i=2}^n \frac{Z_i^2}{\sigma^2}.$$

As $Z_1/\sigma \sim N(||\boldsymbol{\mu}||/\sigma, 1)$ is independent of $Z_i/\sigma \sim N(0, 1)$ for $i = 2, \ldots, n$ which are themselves independent, we have the additional characterization of the noncentral χ^2-distribution as the square of a $N(\delta, 1)$ random variable added to an independent central χ^2 random variable with $n - 1$ df. Hence, the noncentral χ^2-distribution only depends on the parameter δ. □

Finally, our result on the projection of $\mathbf{X}$ follows.

THEOREM 3.2 Let $\mathbf{X} \sim N(\boldsymbol{\mu}, \sigma^2\mathbf{I})$ take values in R^n and $\mathbf{P}$ be a projection onto a subspace of dimension k. Then $||\mathbf{PX}||^2/\sigma^2$ is a noncentral χ^2 random variable with k df and noncentrality parameter $\delta = ||\mathbf{P}\boldsymbol{\mu}||/\sigma$; that is, $||\mathbf{PX}||^2/\sigma^2 \sim \chi_k{}^2(||\mathbf{P}\boldsymbol{\mu}||/\sigma)$.

[2] The noncentrality parameter has other definitions including δ^2. In a particular situation, care should be taken that the definition is understood.

Proof Let $\mathbf{b}_1, \ldots, \mathbf{b}_k, \mathbf{b}_{k+1}, \ldots, \mathbf{b}_n$ be an orthonormal basis for R^n such that the first k vectors are an orthonormal basis for the subspace being projected onto. Changing to the $\mathbf{b}_i$ orthonormal coordinate system, we have

$$\mathbf{Z} = \begin{pmatrix} \mathbf{b}_1'\mathbf{X} \\ \vdots \\ \mathbf{b}_k'\mathbf{X} \\ \mathbf{b}_{k+1}'\mathbf{X} \\ \vdots \\ \mathbf{b}_n'\mathbf{X} \end{pmatrix} \sim N\left(\begin{pmatrix} \mathbf{b}_1'\boldsymbol{\mu} \\ \vdots \\ \mathbf{b}_k'\boldsymbol{\mu} \\ \mathbf{b}_{k+1}'\boldsymbol{\mu} \\ \vdots \\ \mathbf{b}_n'\boldsymbol{\mu} \end{pmatrix}, \sigma^2\mathbf{I}\right).$$

Problem 3.11 shows that the marginal distribution of the first k components of $\mathbf{Z}$ is multinormal, specifically

$$\begin{pmatrix} \mathbf{b}_1'\mathbf{X} \\ \vdots \\ \mathbf{b}_k'\mathbf{X} \end{pmatrix} \sim N\left(\begin{pmatrix} \mathbf{b}_1'\boldsymbol{\mu} \\ \vdots \\ \mathbf{b}_k'\boldsymbol{\mu} \end{pmatrix}, \sigma^2\mathbf{I}^{k\times k}\right).$$

The previous lemma shows that $\sum_{i=1}^{k}[(\mathbf{b}_i'\mathbf{X})^2/\sigma^2]$ is $\chi_k{}^2(\delta)$ with noncentrality parameter $\delta = \sqrt{\sum_{i=1}^{k}(\mathbf{b}_i'\boldsymbol{\mu})^2}/\sigma$. But (Problem 3.4) notice that

$$\sum_{i=1}^{k}\frac{(\mathbf{b}_i'\mathbf{X})^2}{\sigma^2} = \frac{\|\mathbf{PX}\|^2}{\sigma^2} \quad \text{and} \quad \delta = \frac{\sqrt{\|\mathbf{P}\boldsymbol{\mu}\|^2}}{\sigma} = \frac{\|\mathbf{P}\boldsymbol{\mu}\|}{\sigma}. \qquad \square$$

Combining our previous work, we have:

THEOREM 3.3 Let $S_1, \ldots, S_k$ be orthogonal subspaces of R^n whose dimensions add to n; that is, $S_1 + \cdots + S_k = R^n$. Let $\mathbf{P}_i$ be the projection onto S_i. Let $\mathbf{X} \sim N(\boldsymbol{\mu}, \sigma^2\mathbf{I})$ (note $\sigma^2\mathbf{I}$ is the normal covariance matrix); then:

(i) Pythagorean theorem: $\|\mathbf{X}\|^2 = \|\mathbf{P}_1\mathbf{X}\|^2 + \cdots + \|\mathbf{P}_k\mathbf{X}\|^2$.

(ii) The $\|\mathbf{P}_i\mathbf{X}\|^2/\sigma^2$ are *independent* noncentral χ^2 random variables with dimension of S_i df and noncentrality parameters $\delta_i = \|\mathbf{P}_i\boldsymbol{\mu}\|/\sigma$, $i = 1, \ldots, k$.

Proof (i) is given in Problem 3.9. The distributions of the $\|\mathbf{P}_i\mathbf{X}\|^2/\sigma^2$ are given in Theorem 3.2 and the independence follows from Theorem 3.1. $\square$

SUMMARY

The Geometric Approach

The key geometric idea of the analysis of variance is that of projecting the data vector $\mathbf{X}$ onto orthogonal subspaces. For multinormal observations with a covariance matrix of the form $\sigma^2\mathbf{I}$, the projections of the data vector onto orthogonal subspaces are independently distributed multinormal random vectors. If $\mathbf{X}$ takes values in R^n and the dimensions of k orthogonal subspaces add to n, and we let $\mathbf{P}_i$ denote the projection onto subspace S_i

(i) Pythagorean theorem: $\|\mathbf{X}\|^2 = \|\mathbf{P}_1\mathbf{X}\|^2 + \cdots + \|\mathbf{P}_k\mathbf{X}\|^2$.

(ii) The $\|\mathbf{P}_i\mathbf{X}\|^2/\sigma^2$, $i = 1, \ldots, k$, are independent noncentral χ^2 random variables with dimension of S_i df and noncentrality parameters $\delta_i = \|\mathbf{P}_i\boldsymbol{\mu}\|/\sigma$, where $\boldsymbol{\mu} = E(\mathbf{X})$.

Useful Coordinate Systems (Continued)

(3) If we are projecting the data vector onto a subspace of dimension k, let $\mathbf{b}_1, \ldots, \mathbf{b}_k$ be an orthonormal basis for the subspace being projected onto. Extending these vectors to an orthonormal basis, we obtain a coordinate system more convenient for theoretical understanding.

PROBLEMS

3.1 (a) Show that the moment generating function for a χ^2 random variable with r df is

$$M(t) = (1 - 2t)^{-r/2}, \qquad |t| < \tfrac{1}{2}.$$

(b) Show that if X and Y are independent χ^2 random variables with r and s df respectively, then $X + Y$ is a χ^2 random variable with $r + s$ df.

(c) Show that the t variable of equation (1) does have a t-distribution with $n + m - 2$ df.

3.2 Show that Definition 3.1 *uniquely* defines the projection.

3.3 Show that the projection is a linear operator; that is, $\mathbf{P}(a\mathbf{x} + b\mathbf{y}) = a\mathbf{P}\mathbf{x} + b\mathbf{P}\mathbf{y}$.

3.4 Let $\mathbf{b}_1, \ldots, \mathbf{b}_k$ be an orthonormal basis for a subspace S. Show that the projection operator onto S is given by $\mathbf{P} = \sum_{i=1}^{k} \mathbf{b}_i\mathbf{b}_i'$; that is, $\mathbf{P}\mathbf{X} = \sum_{i=1}^{k} \mathbf{b}_i(\mathbf{b}_i'\mathbf{X})$. Also show that $\|\mathbf{P}\mathbf{X}\|^2 = \sum_{i=1}^{k} (\mathbf{b}_i'\mathbf{X})^2$.

3.5 (a) Show that a set of nonzero orthogonal vectors are linearly independent.

(b) Let $S_1, \ldots, S_k$ be orthogonal subspaces of R^n. Then $d_1 + \cdots + d_k \leq n$, where d_i = dimension S_i.

3.6 Let $\mathbf{P}_1$ and $\mathbf{P}_2$ be projections onto orthogonal subspaces. Show that

(a) $\mathbf{P}_i$ is symmetric [i.e., for all $\mathbf{X}$ and $\mathbf{Y}$, $\mathbf{X}'\mathbf{P}_i\mathbf{Y} = (\mathbf{P}_i\mathbf{X})'\mathbf{Y}$] and idempotent; that is, $\mathbf{P}_i^2 = \mathbf{P}_i$, $i = 1, 2$.

(b) $\mathbf{P}_1\mathbf{P}_2 = \mathbf{0}$ ($\mathbf{0}$ is the $\mathbf{0}$ operator; that is, for all $\mathbf{X}$, $\mathbf{0}\mathbf{X} = \mathbf{0}$ the zero vector).

3.7 Show that if $X \sim \chi_n^2(\delta)$, then $E(X) = \delta^2 + n$.

3.8 Let $\mathbf{X}_1, \ldots, \mathbf{X}_k$ be jointly distributed vectors with dimensions $d_1, \ldots, d_k$ respectively. The joint moment generating function (if it exists) is $M(\mathbf{t}_1, \ldots, \mathbf{t}_k) = E(\exp \sum_{i=1}^{k} \mathbf{t}_i'\mathbf{X}_i)$ ($\mathbf{t}_i$ a d_i-dimensional vector).

(a) Assume that the joint moment generating function exists. Show (one-line from Problem 2.4) that the moment generating function determines the joint distribution.

(b) Show (in the case where each $\mathbf{X}_i$ has a probability density function) that the $\mathbf{X}_i$ are independent if and only if

$$M(\mathbf{t}_1, \ldots, \mathbf{t}_k) = M(\mathbf{t}_1, \mathbf{0}, \ldots, \mathbf{0})M(\mathbf{0}, \mathbf{t}_2, \mathbf{0}, \ldots, \mathbf{0}) \cdots M(\mathbf{0}, \ldots, \mathbf{t}_k).$$

(This result holds in general and will be used in its more general form to prove Theorem 3.1.) [*Hint*: Show that $M(\mathbf{t}_1, \mathbf{0}, \ldots, \mathbf{0})$ is the moment generating function of $\mathbf{X}_1$, etc.]

3.9 (a) (*Bessel's inequality*) Let $\mathbf{P}_i$, $i = 1, \ldots, k$, be projections onto orthogonal subspaces. Show for all $\mathbf{X}$ that $\sum_{i=1}^{k} \|\mathbf{P}_i\mathbf{X}\|^2 \leq \|\mathbf{X}\|^2$.

(b) (*Pythagorean theorem and Bessel's equality*) Let $\mathbf{P}_i$, $i = 1, \ldots, k$, be projections onto orthogonal subspaces of R^n. Let the sum of the dimensions of the subspaces equal n. Show for all $\mathbf{X}$ in R^n that $\sum_{i=1}^{k} \|\mathbf{P}_i\mathbf{X}\|^2 = \|\mathbf{X}\|^2$.

3.10 Let $S_1, \ldots, S_k$ be orthogonal subspaces of R^n. The subspace orthogonal to $S_1 + \cdots + S_k$ or the orthogonal complement of $S_1 + \cdots + S_k$ is the subspace consisting of all vectors orthogonal to each S_i, $i = 1, \ldots, k$.

(a) Show that the orthogonal complement of any nonempty set of vectors is a subspace.

(b) If $\dim S_1 + \cdots + \dim S_k = r < n$ and $\mathbf{b}_1, \ldots, \mathbf{b}_{n-r}$ are orthonormal vectors perpendicular to S_i, $i = 1, 2, \ldots, k$, show that

(i) $\{\mathbf{b}_1, \ldots, \mathbf{b}_{n-r}\}$ is a basis for the orthogonal complement of $S_1 + \cdots + S_k$;

(ii) the projection onto the orthogonal complement is given by $\mathbf{PX} = (\mathbf{X}'\mathbf{b}_1)\mathbf{b}_1 + \cdots + (\mathbf{X}'\mathbf{b}_{n-r})\mathbf{b}_{n-r}$.

[Note that to show (i) you must show that $\dim S_1 + \cdots + \dim S_k + \dim(\text{orthogonal complement}) = n$.]

3.11 Let $\mathbf{X}^{n \times 1} \sim N(\boldsymbol{\mu}, \mathbf{C})$, where

$$\mathbf{C} = \begin{array}{c} \\ k\{ \end{array}\!\!\begin{pmatrix} \overbrace{\mathbf{C}_{11}}^{k} & \mathbf{C}_{12} \\ \mathbf{C}_{21} & \mathbf{C}_{22} \end{pmatrix} \quad \text{and} \quad k < n.$$

From the moment generating function of $\mathbf{X}$, find the moment generating function of $\mathbf{Y}$, $\mathbf{Y}' = (X_1, \ldots, X_k)$. From this, conclude that

$$\mathbf{Y} \sim N\left(\begin{pmatrix} \mu_1 \\ \vdots \\ \mu_k \end{pmatrix}, \mathbf{C}_{11}\right).$$

3.12 One thrust of modern statistical theory and application is the concept of *robustness.* Anyone who gives serious thought to the use of statistical models soon concludes that the assumptions of the models used are not often met. For example, in this text normal distributions are extensively used. Usually it is clear that all values from $-\infty$ to $+\infty$ are not possible (e.g., the variable of blood pressure is necessarily a nonnegative number). Also, data only comes with a finite degree of accuracy, meaning that we are observing a discrete random variable (albeit one with a very large number of possible values). Of more concern in practice than the above are *outliers.* Roughly speaking, an outlier has been "defined" in one of two ways: (1) a spuriously extreme value (i.e., a "false" value due to extreme experimental error or malfunction), or (2) a value that may be close to the "true" value but which is so extreme that the one observation may have an undue influence on the value of the statistic.

Returning to the idea of robustness, a statistical test or estimate is *robust* if it works well (i.e., almost according to the theory) when the assumptions are "not quite" satisfied. Thus, a test might be robust for large samples; that is, as n becomes large the theory "holds". A test is robust against outliers if a few extreme values do not unduly influence the test statistic.

Let us consider briefly the robustness of the two-sample t-test against an outlier. Suppose $\bar{x} = 0$, $s_x^2 = .1$, $\bar{y} = -1$, $s_y^2 = .1$, where $n = m$. Do the following for $n = m = 2, 10, 100$:

(a) Calculate the test statistic t.

(b) Suppose one new observation is added to the X observations at the point x_0. Write t as a function of x_0.

(c) Plot $t(x_0)$.

(d) Does t seem robust against outliers?

3.13 (*Robustness continued*) One approach to making a two-sample test that is robust against outliers is to turn to a *rank test.* Let the X *and* Y observations be arranged in nondecreasing order from the smallest observation to the largest. The *rank* of an observation is its position in line; that is, the smallest observation has rank 1, the next smallest has rank 2, up to the largest observation which has rank $m + n$. (Assume there are no ties.) Let T be the sum of the ranks of the Y observations. T is called the *Wilcoxon statistic.*

(a) For the following data, compute the Wilcoxon statistic.

X values: 31.5, 35.1, 32.1, 34.2, 26.7, 31.9, 30.8, 27.3, 27.4, 29.0, 30.0, 36.4, 39.8, 32.0, 35.9, 29.9, 32.2, 31.8.

Y values: 26.5, 22.7, 27.5, 24.9, 23.4.

A closely related statistic is the *Mann–Whitney statistic* U. For all mn pairs of X_i's and Y_j's, let U be the number of pairs (X_i, Y_j) with $Y_j < X_i$.

(b) Compute U for the same data.

(c) U and T are related by the equation $U + T = mn + \frac{1}{2}[m(m + 1)]$. Verify this formula in this case. Tables of critical values for the Wilcoxon (or equivalently, the Mann–Whitney) statistic are in most collections of statistical tables, for example, Owen *et al.*[3]

3.14 If X and Y random samples come from the same continuous distribution, show that the distribution of the Wilcoxon statistic T is the same regardless of which distribution is being sampled (i.e., T is a *distribution-free statistic*).

3.15 Using the data presented in Table 3.1,

(a) compare the control men and women on the six scales;

(b) compare the IM males with the IM females on the six scales.

3.16 Calculate the two-sample t-tests for the data in Table 3.2. What can you say about the p values from the tables in Appendix 2?

3.17 Calculate the two-sample t-tests for the data in Table 3.3. What can you say about the p values from the tables in Appendix 2?

3.18 Define a noncentral t-distribution with n df and noncentrality parameter δ to be the distribution resulting from division of a $N(\delta,1)$ random variable by the square root of an independent $\chi_n{}^2/n$ random variable.

Denote the distribution by $t_n(\delta)$.

(a) What is the distribution of $t_n(0)$?

(b) Show that the distribution of $t_n(-\delta)$ is the same as the distribution of $-t_n(\delta)$.

[3] R. Odeh, D. B. Owen, Z. W. Birnbaum, and L. Fisher, *Pocketbook of Statistical Tables*. Dekker, New York, 1977.

4 THE *k*-SAMPLE COMPARISON OF MEANS (ONE-WAY ANALYSIS OF VARIANCE)

Let us slightly generalize the problem of the previous chapter. Suppose that instead of two independent random samples we now have independent random samples from k ($k \geq 2$) populations. Suppose we model this situation as follows:

$$X_{ij} \sim N(\mu_i, \sigma^2), \qquad i = 1, \ldots, k, \qquad j = 1, \ldots, n_i,$$

all independent random variables; that is, the ith population is sampled n_i times, and it has also been assumed that the variance is the same in each population.

With our current background, the approach to the problem is clear:

(1) Consider the X_{ij}'s as a multinormally distributed random vector $\mathbf{X}$; that is,

$$\mathbf{X} = \begin{pmatrix} X_{11} \\ \vdots \\ X_{1n_1} \\ X_{21} \\ \vdots \\ X_{2n_2} \\ \vdots \\ X_{k1} \\ \vdots \\ X_{kn_k} \end{pmatrix} \sim N(\mu_1 \mathbf{v}_1 + \mu_2 \mathbf{v}_2 + \cdots + \mu_k \mathbf{v}_k, \sigma^2 \mathbf{I}),$$

where $\mathbf{v}_j$ has all zero entries except for ones in positions $(\sum_{m=1}^{j-1} n_m) + 1$ to $\sum_{m=1}^{j} n_m$. Note that $\mathbf{v}_i$ and $\mathbf{v}_j$ are orthogonal for $i \neq j$.

(2) The orthogonality of $\mathbf{v}_1, \ldots, \mathbf{v}_k$ allows independent estimates of the parameters $\mu_1, \ldots, \mu_k$ to be made by projecting the data vector $\mathbf{X}$ onto the orthogonal one-dimensional subspaces generated by $\mathbf{v}_1, \ldots, \mathbf{v}_k$. Thus (can you show this?),

$$\hat{\mu}_i = \frac{\mathbf{v}_i'\mathbf{X}}{\|\mathbf{v}_i\|^2} = \frac{\sum_{j=1}^{n_i} X_{ij}}{n_i} \equiv \bar{X}_{i\cdot},$$

where $\equiv$ denotes "is defined to be."

(3) The $(n-k)$-dimensional subspace orthogonal to $\{\mathbf{v}_1, \ldots, \mathbf{v}_k\}$ may then be used to independently estimate the "nuisance parameter" σ^2. This allows the variability in the estimates $\hat{\mu}_i$, $i = 1, \ldots, k$, to be evaluated. Once again, an unbiased estimate of σ^2 is given by the squared length of the projection of the data vector onto the orthogonal complement of the subspace in which the mean vector is assumed to lie, divided by $n-k$, the dimensionality of the subspace we are projecting onto.

We shall now sneakily use the Pythagorean theorem to find the length of the projection squared without explicitly constructing the projection operator. Let $\mathbf{P}_{k+1}$ be the projection onto the subspace orthogonal to $\{\mathbf{v}_1, \ldots, \mathbf{v}_k\}$. Then we know that if $\mathbf{P}_i$ is the projection onto the $\mathbf{v}_i$ direction,

$$\mathbf{P}_i\mathbf{X} = \frac{\mathbf{v}_i}{\|\mathbf{v}_i\|}\frac{\mathbf{v}_i'\mathbf{X}}{\|\mathbf{v}_i\|} = \bar{X}_{i\cdot}\mathbf{v}_i,$$

so that

$$\|\mathbf{P}_i\mathbf{X}\|^2 = n_i\bar{X}_{i\cdot}^2.$$

Now the Pythagorean theorem says that

$$\|\mathbf{X}\|^2 = \|\mathbf{P}_1\mathbf{X}\|^2 + \cdots + \|\mathbf{P}_k\mathbf{X}\|^2 + \|\mathbf{P}_{k+1}\mathbf{X}\|^2,$$

so that

$$\|\mathbf{P}_{k+1}\mathbf{X}\|^2 = \|\mathbf{X}\|^2 - n_1\bar{X}_{1\cdot}^2 - n_2\bar{X}_{2\cdot}^2 - \cdots - n_k\bar{X}_{k\cdot}^2 = \sum_{i=1}^{k}\sum_{j=1}^{n_i}(X_{ij} - \bar{X}_{i\cdot})^2.$$

Now

$$\frac{\|\mathbf{P}_{k+1}\mathbf{X}\|^2}{\sigma^2} \sim \chi_{n-k}^2\left(\frac{\|\mathbf{P}_{k+1}(\mu_1\mathbf{v}_1 + \cdots + \mu_k\mathbf{v}_k)\|}{\sigma}\right)$$

or

$$\frac{\|\mathbf{P}_{k+1}\mathbf{X}\|^2}{\sigma^2} \sim \chi_{n-k}^2(0),$$

since $\mathbf{P}_{k+1}(\mu_1\mathbf{v}_1 + \cdots + \mu_k\mathbf{v}_k) = \mathbf{0}$. (Why?) Since the mean of a central χ^2 random variable equals its degrees of freedom,

$$E\left(\frac{\|\mathbf{P}_{k+1}\mathbf{X}\|^2}{\sigma^2}\right) = n - k,$$

and therefore

$$\frac{\|\mathbf{P}_{k+1}\mathbf{X}\|^2}{n-k} = \frac{\sum_{i=1}^{k}\sum_{j=1}^{n_i}(X_{ij} - \bar{X}_{i\cdot})^2}{n-k}$$

is an unbiased estimator of σ^2.

Suppose we are interested in testing whether or not all the μ_i's are equal. If all the μ_i are equal, then $\mathbf{X} \sim N(\mu\mathbf{1}, \sigma^2\mathbf{I})$. Note that $\mathbf{1}$ lies in the subspace generated by $\{\mathbf{v}_1, \ldots, \mathbf{v}_k\}$; in fact, $\mathbf{1} = \mathbf{v}_1 + \cdots + \mathbf{v}_k$. Since the Pythagorean theorem holds for a proper subspace of a vector space, we have, using the theorem,

$$\|\mathbf{P}_1\mathbf{X} + \cdots + \mathbf{P}_k\mathbf{X}\|^2 = \|\mathbf{P}_S\mathbf{X}\|^2 + \|\mathbf{P}_T\mathbf{X}\|^2,$$

where S is the subspace spanned by the $\mathbf{1}$ vector and T is the orthogonal complement of S *relative* to the subspace spanned by $\{\mathbf{v}_1, \ldots, \mathbf{v}_k\}$ [a $(k-1)$-dimensional subspace]. If $\mu_i = \mu$ for all i using $\mathbf{P}_T\mathbf{1} = \mathbf{0}$, we have that $\|\mathbf{P}_T\mathbf{X}\|^2/\sigma^2$ has a (central) χ^2-distribution with $k - 1$ df.

Now $\mathbf{P}_S\mathbf{X} = \bar{X}_{\cdot\cdot}\mathbf{1}$, where $\bar{X}_{\cdot\cdot} = \sum_{i=1}^{k}\sum_{j=1}^{n_i} X_{ij}/n$ and $n \equiv \sum_{i=1}^{k} n_i$, so that

$$\|\mathbf{P}_S\mathbf{X}\|^2 = n\bar{X}_{\cdot\cdot}^2.$$

and

$$\|\mathbf{P}_T\mathbf{X}\|^2 = \|\mathbf{P}_1\mathbf{X}\|^2 + \cdots + \|\mathbf{P}_k\mathbf{X}\|^2 - \|\mathbf{P}_S\mathbf{X}\|^2 = n_1\bar{X}_{1\cdot}^2 + \cdots + n_k\bar{X}_{k\cdot}^2 - n\bar{X}_{\cdot\cdot}^2$$

$$= \sum_{i=1}^{k}\sum_{j=1}^{n_i}(\bar{X}_{i\cdot} - \bar{X}_{\cdot\cdot})^2.$$

If the μ_i are not all equal, then $\|\mathbf{P}_T\mathbf{X}\|^2/\sigma^2$ has a noncentral χ^2-distribution with noncentrality parameter $\sqrt{\sum_{i=1}^{k}\sum_{j=1}^{n_i}(\mu_i - \mu)^2}/\sigma$, where $\mu \equiv \sum_{i=1}^{k} n_i\mu_i/n$ (a weighted average of the μ_i's) (Problem 4.4).

If we knew σ^2, $\|\mathbf{P}_T\mathbf{X}\|^2/\sigma^2$ would be a nice statistic to test the hypothesis that $\mu_i = \mu$ for all i. $\|\mathbf{P}_T\mathbf{X}\|^2/\sigma^2$ is not a statistic if σ^2 is unknown, and our "nuisance parameter" is still a nuisance in this case. The trick we shall use to get rid of the dependency on σ^2 is the same "canceling" trick we used in previous chapters:

$$\frac{\|\mathbf{P}_T\mathbf{X}\|^2/\sigma^2}{\|\mathbf{P}_{k+1}\mathbf{X}\|^2/\sigma^2} = \frac{\|\mathbf{P}_T\mathbf{X}\|^2}{\|\mathbf{P}_{k+1}\mathbf{X}\|^2}$$

is the ratio of a noncentral χ^2 to an independent central χ^2 random variable. It will be convenient to divide each χ^2 random variable by the appropriate degrees of freedom.

DEFINITION 4.1 A random variable F has a noncentral F-distribution with noncentrality parameter δ and r numerator and s denominator degrees of freedom, denoted by $F \sim F_{r,s}(\delta)$, if its distribution is the same as that of $(X/r)/(Y/s)$, where $X \sim \chi_r^2(\delta)$, $Y \sim \chi_s^2$, and X and Y are independent random variables.

By (a new) definition we have a *statistic* to investigate the equality of the μ_i, namely,

$$F = \frac{\sum_{i=1}^{k} \sum_{j=1}^{n_i} (\bar{X}_{i\cdot} - \bar{X}_{\cdot\cdot})^2/(k-1)}{\sum_{i=1}^{k} \sum_{j=1}^{n_i} (X_{ij} - \bar{X}_{i\cdot})^2/(n-k)}.$$

Note that if $\mu_i = \mu$ for all i, $\delta = 0$ and F has the usual or central F-distribution.

Let us now turn to what happens when $\delta \neq 0$; that is, the μ_i's differ among themselves. Recall that the expected value of a $\chi_m^2(\delta)$ random variable is $\delta^2 + m$. Then

$$E(F) = \frac{n-k}{k-1}(\delta^2 + k - 1)E\left(\frac{1}{\|\mathbf{P}_{k+1}\mathbf{X}\|^2}\right)$$

which increases as δ^2 increases. Thus, it is plausible that we are more likely to find F too large if δ^2 is larger. (For a precise statement of this result, see Problem 4.2.) The appropriate p value for the data (Problem 4.3) (or the appropriate critical region for testing μ_i all equal) will be found by the probability that a central F statistic is greater than the observed value (or by using a one-sided critical region and rejecting when F is too large). Now consider some examples.

Example 4.1 [N. H. Brodie, R. L. McGhie, H. O'Hara, A. E. P. Rodway, J. C. Valle-Jones, and A. A. Schiff, Once daily administration of a fluphenazine/nortriptyline preparation in the treatment of mixed anxiety/depressive states, *Current Medical Research and Opinion* **4**, 346 (1976)] Pharmacokinetic investigations have shown that the steady state plasma concentration of many drugs is not materially altered whether the drug is taken in divided doses or as a single daily dose. In recent years, it has become evident that once-daily administration is an advantage in that it increases the reliability with which medication is taken by the patient, and several studies have shown that a single dose on retiring is likely to be taken more reliably by patients than several doses during the day. A study of 223 patients diagnosed as suffering from mixed anxiety/depressive states was carried out at five general practice centers to compare the effectiveness of treatment with a once-daily tablet preparation containing 1.5 mg fluphenazine plus 30 mg nortriptyline taken either at night or in the morning, with the same daily dose taken as 1 tablet 3 times a day.

Patients aged 18 to 65 years, who met certain inclusion criteria (which will not be discussed), were randomly allocated to one of three treatment groups:

Group 1 divided dosage (3 times a day),
Group 2 single dose in the morning,
Group 3 single dose at night.

Treatment was continued for 4 weeks and patients were assessed when first seen, after 1 week, and after 4 weeks with a "physician's clinical rating scale" and a "patient's visual analogue scale" the details of which will not be discussed. Of the 223 patients entered in the study, 9 failed to complete the trial period (4 dropouts from Group 1, 4 dropouts from Group 2, and 1 dropout from Group 3).

The results from 214 patients were analyzed for improvements within each group during the treatment period (Day 0 to Day 7, Day 0 to Day 28) using a paired t-test for each rating scale. Each of the treatment groups showed highly significant improvements ($p < .001$) during the course of the study on both physician's and patient's self-rating scales. Since the pretreatment scores of the three groups did not differ significantly, the results from the 214 patients were analyzed for between-group comparison of improvements using the analysis of variance (Tables 4.1 and 4.2). The between-group com-

TABLE 4.1

Between-Group Comparison of Changes in Physician's Clinical Rating Scales: Analysis of Variance[a]

	Day 0 to Day 7				Day 0 to Day 28			
Source	df	SS	MS	F	df	SS	MS	F
Treatments	2	134.38	67.19	5.41[b]	2	453.48	226.74	8.44[c]
Residual	220	2717.84	12.41		211	5668.91	26.87	
Total	222	2852.22			213	6122.39		

[a] F value as given in text; additional notation explained in Chapter 5, page 61.
[b] $p < .01$. [c] $p < .001$.

TABLE 4.2

Between-Group Comparison of Changes in Patient's Visual Analogue Scores: Analysis of Variance

	Day 0 to Day 7				Day 0 to Day 28			
Source	df	SS	MS	F	df	SS	MS	F
Treatments	2	51130	25565	1.956[a]	2	59990	29995	2.964[a]
Residual	212	2771000	13071		202	2044000	10119	
Total	214	2822130			204	2103990		

[a] Not significant.

parison showed significant variation in improvements on the physician's clinical rating scales, which was further analyzed by the Scheffé multiple comparison method to test for pairwise differences between the treatment groups (this method will be discussed in Chapter 8). This showed that the improvement of the morning dosage group was not as great as that of the other two groups. The between-group comparison showed no significant variation in improvements on the patient's self-rating scales.

The substantial reduction in symptom scores which occurred in all treatment groups over the 4-week period confirmed the efficacy of fluphenazine/nortriptyline taken as a single daily dose before retiring at night than on awakening in the morning. However, the size of the actual differences between the mean improvements of each treatment group makes it unlikely that these were of any clinical significance (as opposed to statistical significance). This view finds support in the fact that analysis of the improvements on the patient's self-rating scales failed to reveal significant differences between the groups.

Example 4.2 [J. H. Wilmore, R. B. Parr, W. L. Haskell, D. L. Costill, L. S. Milburn, and R. K. Kerlan, Football pros' strengths—and CV weaknesses—Charted, *The Physician and Sportsmedicine* **4** (Oct 1976)] The physical and physiological status of 185 professional football players was evaluated by position.

One aspect of the study dealt with the cardiovascular (CV) conditioning of the athletes. The players ran on a treadmill to self-determined exhaustion. The oxygen intake in liters per minute was measured. A commonly used measure of endurance is to adjust the oxygen intake by body weight and examine the intake in milliliters per kilogram per minute. The last two columns of Table 4.3 contain the values for the athletes by position.

TABLE 4.3

Maximal Cardiovascular Endurance Capacity of Professional Football Players[a]

Players	No.	HR_{max} (beats/min)	$V_{E\ max}$ (liters/min) BTPS[b]	$\dot{V}_{O_2\ max}$ (liters/min)	$\dot{V}_{O_2\ max}$ (ml/kg/min)
Defensive backs	25	183.2 ± 8.6	143.1 ± 20.3	4.5 ± 0.4	53.1 ± 6.2
Offensive backs, wide receivers	39	180.3 ± 7.7	147.2 ± 20.4	4.7 ± 0.5	52.2 ± 5.0
Linebackers	28	178.7 ± 8.5	159.1 ± 17.0	5.3 ± 0.6	52.1 ± 4.9
Offensive linemen, tight ends	35	184.4 ± 9.0	174.2 ± 30.5	5.6 ± 0.8	49.9 ± 6.6
Defensive linemen	27	183.5 ± 10.6	154.7 ± 24.5	5.3 ± 0.6	44.9 ± 5.4
Quarterbacks,	14	184.1 ± 7.2	148.5 ± 19.9	4.5 ± 0.5	49.0 ± 8.1

[a] Values represent mean ± SD.

[b] Body temperature, ambient pressure, saturated with water.

The authors comment that "While the average college-age male has a $\dot{V}_{O_2\,\max}$ between 48 and 50 ml/kg/min, highly conditioned endurance athletes will have values in the high 70s and 80s." They further comment that "While the absolute $\dot{V}_{O_2\,\max}$ values of these athletes in liters per minute are extremely impressive, the more accurate reflection of endurance capacity is the $\dot{V}_{O_2\,\max}$ value expressed relative to body weight."

Let us perform a one-way analysis of variance to examine whether the $\dot{V}_{O_2\,\max}$ values (in ml/kg/min) appear to differ between the positions at a 5% significance level.

Table 4.3 gives the mean of each group and the within-group variance. To find the overall mean, one computes

$$\bar{X}_{..} = \frac{\sum_{i=1}^{k} n_i \bar{X}_{i\cdot}}{\sum_{i=1}^{k} n_i} = \frac{8466.9}{168} = 50.40.$$

The pooled (i.e., using all data) estimate of the variance is given by

$$\frac{\sum_{i=1}^{k} (n_i - 1)s_i^2}{\sum_{i=1}^{k} (n_i - 1)} = \frac{24 \times (6.2)^2 + 38 \times (5.0)^2 + \cdots + 13 \times (8.1)^2}{24 + 38 + \cdots + 13}$$

$$= \frac{5612.96}{162} = 34.65.$$

The denominator may also be written as

$$\sum_{i,j} (X_{ij} - \bar{X}_{i\cdot})^2.$$

Direct computation gives

$$\sum_{i,j} (\bar{X}_{i\cdot} - \bar{X}_{..})^2 = \sum_{i=1}^{k} n_i(\bar{X}_{i\cdot} - \bar{X}_{..})^2 = 1242.47.$$

From this, our F statistic is

$$F = \frac{1242.47/5}{5612.96/162} = 7.17.$$

The $\alpha = .05$ critical value for a F random variable with 5 and 162 df is approximately 2.25. As $7.17 > 2.25$, one rejects the null hypothesis that all positions have the same $\dot{V}_{O_2\,\max}$ ml/kg/min.

SUMMARY AND GENERALIZATION

(1) While the data vector $\mathbf{X} \sim N(\boldsymbol{\mu}, \sigma^2\mathbf{I})$ takes values in n-dimensional Euclidean space, the mean vector $\boldsymbol{\mu} = E(\mathbf{X})$ is assumed to lie in a k-

dimensional subspace, denoted by M, which is called the *estimation space.* *Least squares* estimates of parameters relating to the mean vector are those values of the parameters which result in a mean vector nearest the observation vector. These estimators are functions of the projection of the data vector onto the subspace M (hence the name estimation space). The orthogonal complement of the estimation space is called the *error space* and will be denoted by $\mathscr{E}$. (Note that R^n is the direct sum of the estimation space and the error space.) Since the expected value of the projection of the data vector onto the error space is the zero vector, that is, $E(\mathbf{P}_{\mathscr{E}}\mathbf{X}) = \mathbf{P}_{\mathscr{E}}E(\mathbf{X}) = \mathbf{0}$, the squared norm of the projection of the data vector onto the error space divided by σ^2 is a (central) χ^2 random variable with the dimension of the error space degrees of freedom; that is, $\|\mathbf{P}_{\mathscr{E}}\mathbf{X}\|^2/\sigma^2 \sim \chi^2_{\dim \mathscr{E}}$. Therefore, $\|\mathbf{P}_{\mathscr{E}}\mathbf{X}\|^2/\dim \mathscr{E}$ is an unbiased estimator of σ^2, which allows us to assess the amount of error in our parameter estimates (hence the name error space). (Also, note that $\mathbf{P}_{\mathscr{E}} = \mathbf{I} - \mathbf{P}_M$.)

(2) The preceding model is called the *full model,* in contrast to the *reduced model* in which we hypothesize that the mean vector lies in a proper subspace of the estimation space, denoted by H, which is called the *hypothesis space.* If the hypothesis is true, that is, the mean vector lies in the hypothesis space, then the squared norm of the projection of the data vector onto the orthogonal complement of the hypothesis space *relative* to the estimation space divided by σ^2 is a (central) χ^2 random variable with $\dim(M \cap H^{\perp})$ df. This projection operator is $\mathbf{P}_M - \mathbf{P}_H$ (Theorem A30). If the reduced model is true, the statistic

$$\frac{\|(\mathbf{P}_M - \mathbf{P}_H)\mathbf{X}\|^2}{\dim(M \cap H^{\perp})} \bigg/ \frac{\|\mathbf{P}_{\mathscr{E}}\mathbf{X}\|^2}{\dim \mathscr{E}}$$

has a (central) F-distribution with $\dim(M \cap H^{\perp})$ numerator degrees of freedom and $\dim \mathscr{E}$ denominator degrees of freedom and may be used to test the hypothesis.

PROBLEMS

4.1 Show that $\mathbf{P}_1 + \cdots + \mathbf{P}_k$ is the projection onto the subspace spanned by $\{\mathbf{v}_1, \ldots, \mathbf{v}_k\}$. Also show that (Pythagoras again)

$$\|\mathbf{P}_1\boldsymbol{\mu} + \cdots + \mathbf{P}_k\boldsymbol{\mu}\|^2 = \|\mathbf{P}_1\boldsymbol{\mu}\|^2 + \cdots + \|\mathbf{P}_k\boldsymbol{\mu}\|^2$$

when $\boldsymbol{\mu} = \mu_1\mathbf{v}_1 + \cdots + \mu_k\mathbf{v}_k$. See pages 36–37 for notation.

4.2 A random variable X is stochastically less than or equal to a random variable Y if for each x, $P(X \leq x) \geq P(Y \leq x)$. We shall denote this by $X \overset{S}{\leq} Y$.

(a) Let $X \sim \chi_n{}^2(\delta_1)$ and $Y \sim \chi_n{}^2(\delta_2)$. Show that $X \overset{S}{\leq} Y$ if and only if $\delta_1 \leq \delta_2$.

(b) Let $X \sim F_{n,m}(\delta_1)$ and $Y \sim F_{n,m}(\delta_2)$. Show that $X \overset{S}{\leq} Y$ if and only if $\delta_1 \leq \delta_2$.

[*Hint*: Why can you prove (a) with $X = (\delta_1 + X_1)^2 + \sum_{i=2}^n X_i{}^2$, $Y = (\delta_2 + X_1)^2 + \sum_{i=2}^n X_i{}^2$, and $X_i \sim N(0, 1)$ independent, $i = 1, \ldots, n$?]

4.3 One of (many) drawbacks to using a hypothesis testing orientation in science is that the choice of the significance level is somewhat arbitrary, although .10, .05, and .01 have been hallowed by age and tradition. In this context we pass along an unverified bit of gossip:

One of the enjoyable aspects of statistics (as of much of human endeavor) is that it takes place through humans with all the unpredictability that this implies. An example of this was the ongoing feud between two giants of statistics: Sir R. A. Fisher and Karl Pearson. See for example *Statistical Methods and Scientific Inference* written by Fisher after Pearson's demise. At any rate, the apocryphal story is told that when Fisher was writing *Statistical Methods for Research Workers*, the tables he needed had been computed by Pearson's laboratory. Upon trying to get permission to use the tables he was refused, but found out that selected portions could be used without violating the copyright laws. He selected portions that gave the (now standard) critical values, .10, .05, and .01. Fisher's eminence and contributions to statistics were such that these values are standard to this day.

One author of this text has had to deal with more than one frustrated party who had missed "significance" at a .05 level (but achieved it at a .06 level.) In one case, the investigator more or less demanded that an "outlier" that had been removed be reinserted to reestablish "significance."

At any rate, the point is that the selection of a critical value involves a subjective element (as does almost all statistics). A more reasonable approach is that of the p value (p for probability). In a hypothesis testing situation, the p value is a value such that the null hypothesis would be "rejected" if the significance level were greater than this value and accepted if the significance level were less than the value.

Suppose that a one-way analysis of variance for equality of three means based on samples of sizes 11, 14, and 83 respectively gives a value of $F = 5.39$. What inequalities for p can you get from the tables in Appendix 2?

4.4 Show that $\|\mathbf{P}_T\boldsymbol{\mu}\| = \sqrt{\sum_{i=1}^k \sum_{j=1}^{n_i} (\mu_i - \mu)^2}$, where $\mu = \sum_{i=1}^k n_i\mu_i/n$.

4.5 Prove the following:

THEOREM Let C_i, $i = 1, \ldots, p$, be independent noncentral χ^2 random variables with d_i df and δ_i the respective noncentrality parameters. Then $C = C_1 + \cdots + C_p$ is a noncentral χ^2 random variable with $d_1 + \cdots + d_p$ df and $\delta = \sqrt{\delta_1{}^2 + \cdots + \delta_p{}^2}$ as a noncentrality parameter.

4.6 The title of this text is *Fixed Effects Analysis of Variance*. In the fixed effects analysis of variance, the discrete factor that defines the groups contains all groups of interest. In some situations, however, k groups are chosen at random from a much larger possible number of groups. In this case, we wish to extend inferences to the larger population of potential groups. An example of this might be a comparison of student achievement at schools in a state. Schools (or groups) might be selected at random and tests given to n_i students in the ith school.

One model for this situation is

$$Y_{ij} = \mu + \tau_i + e_{ij} \qquad \text{where} \quad \tau_i \sim N(0, \sigma_\tau^2)$$

and $e_{ij} \sim N(0, \sigma^2)$ and all the τ_i's and e_{ij}'s are mutually independent. In this case, to test a group effect is to test $\sigma_\tau^2 = 0$, and the model is said to be a *random effects* model or a variance components model.

(a) Under this random effects model, find $E(\sum_i (\bar{Y}_{i\cdot} - \bar{Y}_{\cdot\cdot})^2/(k-1))$ and $E(\sum_i \sum_j (Y_{ij} - \bar{Y}_{i\cdot})^2/(n-k))$.

(b) Under the null hypothesis of $\sigma_\tau^2 = 0$, what is the distribution of

$$F = \frac{n-k}{k-1} \frac{\sum_i (\bar{Y}_{i\cdot} - \bar{Y}_{\cdot\cdot})^2}{\sum_i \sum_j (Y_{ij} - \bar{Y}_{i\cdot})^2} ?$$

(c) The *intraclass correlation coefficient* is

$$\rho = \frac{\operatorname{cov}(Y_{ij}, Y_{ij'})}{\operatorname{var}(Y_{ij})} = \frac{\sigma_\tau^2}{(\sigma_\tau^2 + \sigma^2)}.$$

Find an unbiased estimate of σ_τ^2, given an estimate of ρ. [*Hint*: Use the results of (a).]

4.7 [P. M. Chikos, M. M. Figley, and L. Fisher, Visual assessment of total heart volume and specific chamber size from standard chest radiographs, *American Journal of Roentgenology* **128**, 375–380 (1977)] Error in the reading of chest x rays was studied. The opinion of 10 radiologists as to whether the left ventricle was normal was compared to data obtained by ventriculography (inserting a catheter into the left ventricle and injecting a radiopaque fluid into the heart and taking a series of x rays). The ventriculography data was used to classify the subject's left ventricle as normal or abnormal. Using this "gold standard," the percentage of errors for each radiologist was computed. The authors were interested in the effect of experience. The radiologists were classified into one of three groups: senior staff, junior staff, and residents. The data for the three groups are given in Table 4.4. Compute the F statistic for testing equality of the means in the three groups. Do you reject the null hypothesis that $\mu_1 = \mu_2 = \mu_3$ at the .05 significance level?

TABLE 4.4

	Senior staff	Junior staff	Residents
i	1	2	3
n_i	2	4	4
x_{ij}	7.3	13.3	14.7
(% error)	7.4	10.6	23.0
		15.0	22.7
		20.7	26.6

4.8 (*Robustness continued*) In Problem 3.13 a rank test for the two-sample problem was given. Kruskal and Wallis introduced an analogous rank test for the one-way analysis of variance situation. Jointly rank the observations from all k groups. Let R_i be the sum of the ranks of the observations from the ith group. The *Kruskal–Wallis statistic* is

$$H = \frac{12}{n(n+1)}\left[\sum_{i=1}^{k} (R_i^2/n_i)\right] - 3(n+1).$$

For the data of Problem 4.7, compute H. If you have access to appropriate tables, do you reject the null hypothesis of all k samples from the same distribution at the .05 significance level?

4.9 [D. Shohori, I. Gedalia, A. E. Nizel, and V. Westreich, Fluoride uptake in rats given tea with milk, *Journal of Dental Research* **55**, No. 5 (Sept–Oct 1976)] Shohori *et al.* studied the fluoride uptake in rats given various drinking fluids. Fluoride uptake may be increased, decreased, or left unchanged, depending on its fluid medium.

A total of 75 male rats were assigned to five groups. Each group was composed of 15 rats and was supplied with one of the following drinking fluids:

Group 1 tea diluted with distilled water containing 2.9 parts per million (ppm) fluoride,
Group 2 tea diluted with milk containing 2.9 ppm fluoride,
Group 3 distilled water containing 0 ppm fluoride,
Group 4 distilled water diluted with milk containing about .034 ppm fluoride,
Group 5 distilled water diluted with milk containing 2.9 ppm fluoride.

A one-to-three dilution ratio was used. The fluids were sweetened with 100 g sugar per liter of fluid to increase consumption and consequently fluoride incorporation. Each rat consumed about the same amount of fluid daily and received a standard diet of up to 5 ppm fluoride ad libitum.

TABLE 4.5

Fluoride Concentration in Molar Surfaces and Femur Ash

	Mean fluoride (ppm)[a]		
	Molar surfaces		
Group	Mandibular	Maxillary	Femur ash
1	35.3 ± 17	25.1 ± 10.5	383 ± 45
2	27.1 ± 21	20.1 ± 6.8	395 ± 56
3	20.3 ± 7.2	17.6 ± 4.5	232 ± 26
4	22.5 ± 8.8	16.6 ± 5.6	205 ± 23
5	18.9 ± 6.0	17.2 ± 4.9	377 ± 100

[a] ±SD.

The rats were killed after 40 days and the fluoride concentration in molar surfaces and femur ash were measured. The mean concentrations for each group ($n = 15$) are presented in Table 4.5.

For the femur ash variable, find the F value for the one-way ANOVA for all five groups. Is this significant at the .05 level?

4.10 [R. Anderson and R. Boggs, Glucogenic and ketogenic capacities of lard, safflower oil, and triundecanoin in fasting rats, *Journal of Nutrition* **105**, 185–189 (1975)] As part of a study on fatty acid oxidation, rats were fed diets containing 30% of either lard (a ketogenic fat), triundecanoin (a glucogenic fat), or safflower oil (a fat high in linoleic acid). Weight gains and feed efficiencies (g of gain/100 g of diet ingested) during weeks 5–8 of the study are presented in Table 4.6. Calculate the F values for the three response variables presented in the table. Values with different superscripts are significantly different from each other at $p < .05$ by Tukey multiple comparison methods (which will be discussed in Chapter 8).

TABLE 4.6

Effects of Dietary Oils on Final Body Weight, Weight Gain, and Feed Efficiency (mean ± SEM for 20 rats per diet)

Dietary oil	Final body weight (g)	Gain	Efficiency
Triundecanoin	$141^a \pm 4$	$46^a \pm 2$	$32.8^a \pm 1.0$
Safflower oil	$195^b \pm 4$	$77^b \pm 3$	$40.2^b \pm 1.4$
Lard	$222^c \pm 5$	$81^b \pm 3$	$41.5^b \pm 1.4$

4.11 Using Table 4.3 concerning the professional football players, test at the $\alpha = .10$ significance level the null hypothesis that the maximal heart rate is the same for all positions.

4.12 Let $E(\mathbf{X}) = \boldsymbol{\mu}$ and $\mathrm{cov}(\mathbf{X}) = \sigma^2\mathbf{I}$ and $\mathbf{P}_S$ be the projection onto a subspace S of dimension k, show that $E(\|\mathbf{P}_S\mathbf{X}\|^2) = \sigma^2 k + \|\mathbf{P}_S\boldsymbol{\mu}\|^2$. Hence, given the above assumptions, conclude that an unbiased estimate of σ^2 is given by the squared length of the projection of the data vector onto the orthogonal complement of the subspace in which the mean vector is assumed to lie, divided by the dimensionality of the orthogonal complement. Thus the assumption that $\mathbf{X}$ is multinormally distributed is *not* needed to obtain an unbiased estimate of σ^2.

5 THE BALANCED TWO-WAY FACTORIAL DESIGN WITHOUT INTERACTION

Consider a situation in which it is desired to measure the response on a continuous variable to combinations of levels of two factors. For example, we might desire to investigate blood pressure for differing sex and blood type distributions or to investigate weight loss comparing three diets among urban, suburban, and rural subjects or to investigate kilometers per liter among 10 types of cars for both city and country driving.

As a first attempt to model this type of situation, let us suppose that each of the two factors has an effect and the effects are additive; that is, we model the response or outcome variable by

$$X_{ij} = \alpha_i + \beta_j$$

when the first factor is at the ith level (number the possible states of the factor) and the second factor is at the jth level. In the following we shall suppose that the first factor has I levels or possibilities or states and the second factor has J possibilities.

There are several weaknesses apparent in this model.

(1) If we repeat a measurement at the same levels i and j, a different value (e.g., blood pressure, weight loss, or kilometers per liter) may result. This, of course, is a happy circumstance in that it implies some statistical analysis will be appropriate and thus the model belongs in this text. Suppose we have K measurements at each combination of levels for the two factors (called the balanced case because each combination has the same number of observations) and assume a normal "error" term:

$$X_{ijk} = \alpha_i + \beta_j + e_{ijk}, \qquad i = 1, \ldots, I, \quad j = 1, \ldots, J, \quad k = 1, \ldots, K,$$

where the e_{ijk} are independent, identically distributed (i.i.d.) $N(0, \sigma^2)$ random variables. (In a later chapter we shall further discuss the assumptions of our models.)

(2) Another problem is that the α_i's and β_j's *cannot be estimated* and thus are not "real" parameters. To see this, consider any constant c; then

$$\alpha_i + \beta_j = (\alpha_i - c) + (\beta_j + c) = \tilde{\alpha}_i + \tilde{\beta}_j,$$

where $\tilde{\alpha}_i \equiv \alpha_i - c$ and $\tilde{\beta}_j \equiv \beta_j + c$. From the observations, there is no way of distinguishing between the α_i,β_j's and the $\tilde{\alpha}_i$,$\tilde{\beta}_j$'s. If several sets of parameters lead to the same distribution for the observations, the model is said to be nonidentifiable. We shall surmount this problem by definition! (Also see the Problems.) That is, restrictions will be placed on the allowable α_i's and β_j's that make them unique. We shall introduce a new parameter μ to be the overall level of response and let the α_i and β_j give the excursions or deviations from the overall mean. Our model is

$$\begin{gathered} X_{ijk} = \mu + \alpha_i + \beta_j + e_{ijk}, \qquad i = 1, \ldots, I, \quad j = 1, \ldots, J, \quad k = 1, \ldots, K; \\ e_{ijk} \quad \text{i.i.d.} \quad N(0, \sigma^2) \qquad \text{and} \qquad \sum_{i=1}^{I} \alpha_i = \sum_{j=1}^{J} \beta_j = 0. \end{gathered} \tag{1}$$

Suppose we wish to investigate whether or not changing the levels of the first factor has any effect; that is, is $\alpha_i = 0$ for all i? Suppose we lexicographically order the X_{ijk} to give us a multinormal model:

$$\mathbf{X} = \begin{pmatrix} X_{111} \\ X_{112} \\ \vdots \\ X_{11K} \\ X_{121} \\ \vdots \\ X_{IJK} \end{pmatrix} \sim N(\mu\mathbf{1} + \alpha_1\mathbf{v}_1 + \cdots + \alpha_I\mathbf{v}_I + \beta_1\mathbf{w}_1 + \cdots + \beta_J\mathbf{w}_J, \sigma^2\mathbf{I}),$$

where $\mathbf{v}_{i_0}$ has all zero entries except for ones corresponding to X_{i_0jk} for all j and k, and similarly for the $\mathbf{w}_j$'s. We might be tempted to think that we could examine the α_i effect by projecting onto the subspace spanned by $\{\mathbf{v}_1, \ldots, \mathbf{v}_I\}$. The trouble with this approach is that it allows all combinations of $\alpha_1, \ldots, \alpha_I$—not only those which sum to zero. In fact, $\sum_{i=1}^{I} \alpha_i = 0$ restricts (Problems 5.2 and 5.3) the appropriate subspace to an $(I - 1)$-dimensional subspace. Since $\alpha_I = -\alpha_1 - \alpha_2 - \cdots - \alpha_{I-1}$ and $\beta_J = -\beta_1 - \cdots - \beta_{J-1}$, the model is

$$\begin{aligned} \mathbf{X} \sim N(\mu\mathbf{1} &+ \alpha_1(\mathbf{v}_1 - \mathbf{v}_I) + \cdots + \alpha_{I-1}(\mathbf{v}_{I-1} - \mathbf{v}_I) + \beta_1(\mathbf{w}_1 - \mathbf{w}_J) + \cdots \\ &+ \beta_{J-1}(\mathbf{w}_{J-1} - \mathbf{w}_J), \sigma^2\mathbf{I}). \end{aligned} \tag{2}$$

Note that $\mathbf{1}'(\mathbf{v}_{i_0} - \mathbf{v}_I) = JK - JK = 0$. In a similar fashion we can show that the three subspaces spanned by $\{\mathbf{1}\}$, $\{\mathbf{v}_1 - \mathbf{v}_I, \ldots, \mathbf{v}_{I-1} - \mathbf{v}_I\}$, and $\{\mathbf{w}_1 - \mathbf{w}_J, \ldots, \mathbf{w}_{J-1} - \mathbf{w}_J\}$ are mutually orthogonal. Thus we can get at the α_i effect by projecting onto the appropriate subspace. In fact, we can get independent assessments of the overall mean, α_i and β_j contributions due to the orthogonality of the subspaces. We proceed to investigate the projections and squared norms in order to evaluate the appropriate χ^2 statistics. Let us start with the overall or grand mean projection.

(1) *The grand mean* μ

$$\text{projection} = \frac{\mathbf{1}}{\|\mathbf{1}\|}\frac{\mathbf{1}'\mathbf{X}}{\|\mathbf{1}\|} = \frac{\sum_i \sum_j \sum_k X_{ijk}}{IJK} \cdot \mathbf{1} \equiv \bar{X}_{\ldots}\mathbf{1},$$

$$\|\text{projection}\|^2 = IJK\bar{X}^2_{\ldots}$$

(2) *The* α *subspace* Let $\mathbf{u}_i = \mathbf{v}_i - \mathbf{v}_I$, $i = 1, \ldots, I-1$. Let the projection of $\mathbf{X}$ onto the α subspace be $\mathbf{Y} = \hat{\alpha}_1\mathbf{u}_1 + \cdots + \hat{\alpha}_{I-1}\mathbf{u}_{I-1}$. Recall (Problem 5.5) that $\mathbf{Y}$ is characterized by $\mathbf{u}_i'\mathbf{Y} = \mathbf{u}_i'\mathbf{X}$, $i = 1, \ldots, I-1$; that is,

$$\begin{aligned} \hat{\alpha}_1\mathbf{u}_1'\mathbf{u}_1 + \cdots + \hat{\alpha}_{I-1}\mathbf{u}_1'\mathbf{u}_{I-1} &= \mathbf{u}_1'\mathbf{X}, \\ &\vdots \\ \hat{\alpha}_1\mathbf{u}_{I-1}'\mathbf{u}_1 + \cdots + \hat{\alpha}_{I-1}\mathbf{u}_{I-1}'\mathbf{u}_{I-1} &= \mathbf{u}_{I-1}'\mathbf{X}. \end{aligned}$$

If we rewrite this in vector form, we see

$$\begin{pmatrix} \mathbf{u}_1' \\ \vdots \\ \mathbf{u}_{I-1}' \end{pmatrix} (\mathbf{u}_1, \ldots, \mathbf{u}_{I-1}) \begin{pmatrix} \hat{\alpha}_1 \\ \vdots \\ \hat{\alpha}_{I-1} \end{pmatrix} = \begin{pmatrix} \mathbf{u}_1' \\ \vdots \\ \mathbf{u}_{I-1}' \end{pmatrix} \mathbf{X}.$$

Note that the dimensions of the matrices (from left to right) are $(I-1) \times IJK$, $IJK \times (I-1)$, $(I-1) \times 1$, $(I-1) \times IJK$, and $IJK \times 1$. The coefficients of the $\mathbf{u}_i$'s and estimates of the α_i's (why?) are given by

$$\begin{pmatrix} \hat{\alpha}_1 \\ \vdots \\ \hat{\alpha}_{I-1} \end{pmatrix} = \left[\begin{pmatrix} \mathbf{u}_1' \\ \vdots \\ \mathbf{u}_{I-1}' \end{pmatrix} (\mathbf{u}_1, \ldots, \mathbf{u}_{I-1})\right]^{-1} \begin{pmatrix} \mathbf{u}_1' \\ \vdots \\ \mathbf{u}_{I-1}' \end{pmatrix} \mathbf{X}.$$

It is straightforward to check that

$$\begin{pmatrix} \mathbf{u}_1' \\ \vdots \\ \mathbf{u}_{I-1}' \end{pmatrix} (\mathbf{u}_1, \ldots, \mathbf{u}_{I-1}) = \begin{pmatrix} 2JK & JK & \ldots & JK \\ JK & 2JK & \ddots & \vdots \\ \vdots & \ddots & \ddots & JK \\ JK & \ldots & JK & 2JK \end{pmatrix}. \tag{3}$$

The inverse has $(I-1)/IJK$ for the diagonal and $-1/IJK$ for the off-diagonal elements. Noting that

$$\mathbf{u}_i'\mathbf{X} = \sum_{j,k} X_{ijk} - \sum_{j,k} X_{Ijk} = JK(\bar{X}_{i\cdot\cdot} - \bar{X}_{I\cdot\cdot}),$$

we find that

$$\begin{aligned}\hat{\alpha}_i &= \frac{(I-1)}{IJK}(JK(\bar{X}_{i\cdot\cdot} - \bar{X}_{I\cdot\cdot})) - \frac{1}{IJK}\sum_{i'\neq i, i'=1}^{I-1} JK\bar{X}_{i'\cdot\cdot} + \frac{(I-2)}{IJK}(JK\bar{X}_{I\cdot\cdot}) \\ &= \bar{X}_{i\cdot\cdot} - \bar{X}_{\cdot\cdot\cdot}\end{aligned}$$

Thus the projection of $\mathbf{X}$ onto the subspace is

$$\hat{\alpha}_1(\mathbf{v}_1 - \mathbf{v}_I) + \cdots + \hat{\alpha}_{I-1}(\mathbf{v}_{I-1} - \mathbf{v}_I) = \hat{\alpha}_1\mathbf{v}_1 + \cdots + \hat{\alpha}_I\mathbf{v}_I,$$

where $\hat{\alpha}_I$ is defined to be $\hat{\alpha}_I = \bar{X}_{I\cdot\cdot} - \bar{X}_{\cdot\cdot\cdot}$. Since the vectors $\mathbf{v}_1, \ldots, \mathbf{v}_I$ are orthogonal,

$$\|\text{projection}\|^2 = \sum_{i=1}^{I} \hat{\alpha}_i^2\|\mathbf{v}_i\|^2 = \sum_{i=1}^{J} JK(\bar{X}_{i\cdot\cdot} - \bar{X}_{\cdot\cdot\cdot})^2.$$

(3) Similarly, *the projection onto the* β *subspace* leads to

$$\hat{\beta}_j = \bar{X}_{\cdot j\cdot} - \bar{X}_{\cdot\cdot\cdot} \qquad \text{and} \qquad \|\text{projection}\|^2 = \sum_{j=1}^{I} IK(\bar{X}_{\cdot j\cdot} - \bar{X}_{\cdot\cdot\cdot})^2.$$

It is not surprising that $\hat{\alpha}_i = \bar{X}_{i\cdot\cdot} - \bar{X}_{\cdot\cdot\cdot}$ since α_i is the average value of the ith level measured from the overall mean μ. It is easy to show $E(\hat{\alpha}_i) = \alpha_i$, $E(\hat{\beta}_j) = \beta_j$, and $E(\hat{\mu}) = \mu$, where $\hat{\mu} = \bar{X}_{\cdot\cdot\cdot}$.

(4) *The Pythagorean theorem* may be used to find the $\|\text{projection}\|^2$ for the error subspace which when divided by the dimensionality of the error space gives the estimate of σ^2, the error variance. It is

$$\begin{aligned}&\|\mathbf{X}\|^2 - \sum_{i=1}^{I} JK(\bar{X}_{i\cdot\cdot} - \bar{X}_{\cdot\cdot\cdot})^2 - \sum_{j=1}^{J} IK(\bar{X}_{\cdot j\cdot} - \bar{X}_{\cdot\cdot\cdot})^2 - IJK\bar{X}^2_{\cdot\cdot\cdot} \\ &\quad = \sum_{i,j,k}(X_{ijk} - \bar{X}_{i\cdot\cdot} - \bar{X}_{\cdot j\cdot} + \bar{X}_{\cdot\cdot\cdot})^2.\end{aligned}$$

Note that the unknown error e_{ijk} can be estimated by

$$\hat{e}_{ijk} \equiv X_{ijk} - (\hat{\mu} + \hat{\alpha}_i + \hat{\beta}_j) = X_{ijk} - \bar{X}_{i\cdot\cdot} - \bar{X}_{\cdot j\cdot} + \bar{X}_{\cdot\cdot\cdot},$$

and that $E(\hat{e}_{ijk}) = 0$. (Why?)

We shall label the factor associated with the α_i's as factor A, and the β factor as factor B. In this problem the Pythagorean theorem has given

$$\begin{aligned}\sum_{i,j,k} X^2_{ijk} &= IJK\bar{X}^2_{\cdot\cdot\cdot} + \sum_i JK(\bar{X}_{i\cdot\cdot} - \bar{X}_{\cdot\cdot\cdot})^2 \\ &\quad + \sum_j IK(\bar{X}_{\cdot j\cdot} - \bar{X}_{\cdot\cdot\cdot})^2 + \sum_{i,j,k}(X_{ijk} - \bar{X}_{i\cdot\cdot} - \bar{X}_{\cdot j\cdot} + \bar{X}_{\cdot\cdot\cdot})^2. \qquad (4)\end{aligned}$$

Viewing equation (4), we can easily see why the analysis of variance is a particularly apt name for the subject we are studying. Notice how we have decomposed the variability of the X_{ijk} into components due to the grand mean, factor A, factor B, and an error (or unaccounted for variability) term.

It is customary and useful to summarize this decomposition in an analysis of variance (ANOVA) table (Table 5.1). The format differs slightly in differing presentations but is basically the same. We first present the table and then discuss it.

TABLE 5.1

ANOVA Table

Source	SS	df	MS	E(MS)
Factor A	$SS_A = \sum_i JK(\bar{X}_{i..} - \bar{X}_{...})^2$	$I-1$	$\frac{SS_A}{I-1} \equiv MS_A$	$\sigma^2 + \frac{JK\sum_i \alpha_i^2}{I-1}$
Factor B	$SS_B = \sum_j IK(\bar{X}_{.j.} - \bar{X}_{...})^2$	$J-1$	$\frac{SS_B}{J-1} \equiv MS_B$	$\sigma^2 + \frac{IK\sum_j \beta_j^2}{J-1}$
Error	$SS_e = \sum_{i,j,k} (X_{ijk} - \bar{X}_{i..} - \bar{X}_{.j.} + \bar{X}_{...})^2$	$IJK - I - J + 1$	$\frac{SS_e}{IJK - I - J + 1} \equiv MS_e$	σ^2
Grand mean	$IJK\bar{X}_{...}^2$	1	$IJK\bar{X}_{...}^2 \equiv MS_{GM}$	$\sigma^2 + IJK\mu^2$
Total	$\sum_{i,j,k} X_{ijk}^2$	IJK		

The source column gives the interpretation of the subspace being considered. In applications, the labeling will reflect the quantity being considered, e.g., sex, educational level.

The $\|\text{projection}\|^2$ terms all are second-order polynomials in the X_{ijk}'s and are called sums of squares (SS). Each SS term is the length of an appropriate projection squared. Note that each sum of squares can be written as the sum over all observations of the square of the least squares estimate of the parameter corresponding to that effect. For example, the sum of squares for factor A can be written $\sum_i \sum_j \sum_k \hat{\alpha}_i^2$; that is, we square our estimate of α_i and sum over all observations.

The degrees of freedom refers to the degrees of freedom of the noncentral χ^2-distribution of the corresponding sum of squares divided by σ^2. This column might just as well be labeled dimension—for the number is the dimension of the subspace onto which the data vector is being projected.

The mean square (MS) "adjusts" the size of the SS term for the number of dimensions involved. The ratio of two mean square terms are used in forming a noncentral F random variable.

The expected mean square column indicates the noncentrality parameter squared times σ^2/df plus σ^2 and clearly indicates which quantities to use to examine which hypothesis. Recall that it is implicit in the ANOVA table that the SS terms divided by σ^2 are independent noncentral χ^2 random variables, so that we may form the statistics given in Table 5.2 to test the various hypotheses.

TABLE 5.2

Hypothesis	Statistic	Distribution
$\mu = 0$	$\frac{MS_{GM}}{MS_e}$	$\sim F_{1,IJK-I-J+1}\left(\frac{\sqrt{IJK\mu^2}}{\sigma}\right)$
$\alpha_i = 0$ (for all i)	$\frac{MS_A}{MS_e}$	$\sim F_{I-1,IJK-I-J+1}\left(\frac{\sqrt{JK\sum_i \alpha_i^2}}{\sigma}\right)$
$\beta_j = 0$ (for all j)	$\frac{MS_B}{MS_e}$	$\sim F_{J-1,IJK-I-J+1}\left(\frac{\sqrt{IK\sum_j \beta_j^2}}{\sigma}\right)$
$\alpha_i = 0$ and $\beta_j = 0$ (for all i and j)	$\frac{(SS_A + SS_B)}{MS_e(I + J - 2)}$	$\sim F_{I+J-2,IJK-I-J+1}\left(\frac{\sqrt{JK\sum_i \alpha_i^2 + IK\sum_j \beta_j^2}}{\sigma}\right)$

Example 5.1 Analyses of variance usually result from planned experiments. Often, variability among experimental units may mask or obscure treatment effects of interest to an experimenter. This nuisance factor of experimental unit differences can be minimized by the use of a randomized block design. Blocking is an experimental design technique using experimental control to reduce variability due to experimental error and to obtain unbiased estimates of treatment effects. A randomized block design is based on the principle of assigning the experimental units to blocks so that the variability among units within any block is less than the variability among the blocks, or in other words, so that the experimental units within each block are as homogeneous as possible. The key to success lies in the experimenter's ability to anticipate which experimental units would respond alike if treated alike. Thus, for example, if there are factors that are known to affect the outcome or response, often the experimental units are put into blocks, where the blocks have similar values on the factor known to affect the outcome. Another example is the use of littermates to form a block since the genetic variability among littermates is less than the genetic variability among the litters. If there are I treatments to be compared, we group the experimental units into blocks of size I and then randomly assign the I treatments to the I experimental units within each block (so that each treatment occurs once in each block). Thus, all treatment comparisons are then made within sets of similar experimental units so that the effect of variations between blocks is eliminated so far as treatment comparisons are concerned.

In the case of a complete randomized block design, often called a randomized block design, the IJ experimental units are divided into J blocks of I experimental units each, and then the I different treatments are randomly assigned to the I units in each block. Our model is

$$X_{ij} = \mu + \alpha_i + \beta_j + e_{ij}, \qquad i = 1, \ldots, I, \quad j = 1, \ldots, J;$$
$$\sum_{i=1}^{I} \alpha_i = \sum_{j=1}^{J} \beta_j = 0, \qquad e_{ij} \text{ i.i.d. } N(0, \sigma^2).$$

Example 5.2 [M. Shermer and E. Perkins, Utilization of heated soybean protein isolate as a dietary source of methionine for rats, *Journal of Nutrition* **104**, 1389–1395 (1974)] Several investigators have observed decreases in protein digestibility that resulted from thermal processing of protein foods and have shown that changes occurred in the availability of some amino acids. Shermer and Perkins studied the effect of different heat treatments of the dietary protein of young rats on the sulfur-containing free amino acids in the plasma. Since other authors have stated that a deficiency of methionine in the diet reduces the level of plasma-free methionine and cystine and since it was shown previously that methionine was partially destroyed under different conditions of heating, one group of diets was not supplemented with methionine (Group I), one was supplemented before the heat treatment (Group II), and one was supplemented after the heat treatment (Group III).

A complete randomized block design was employed in order to differentiate between the heat-treatment effect and the effect of the supplementation of the diet with methionine. The six different heat treatments will not be described but only labeled 1 to 6. The plasma-free methionine level in rats fed heated soybean protein is presented in Table 5.3 in which each

TABLE 5.3

Plasma-Free Methionine Level in Rats Fed Heated Soybean Protein (μmoles/100 ml, Averages for Four Rats)

	Heat treatment					
Group	1	2	3	4	5	6
I (block I)	4.6	5.7	4.9	4.2	4.9	4.9
II (block II)	5.2	5.9	5.8	5.5	5.4	5.6
III (block III)	5.3	5.9	6.0	5.4	5.3	5.6

Two-way analysis of variance

Source	SS	df	MS	F	p Value
Among treatments	1.51	5	.302	5.97	<.01
Between blocks	2.01	2	1.00	19.86	<.001
Experimental error	0.51	10	.051		

observation is the average for four rats (thus each experimental unit consists of four rats) along with the ANOVA table. The different heat treatments of the protein had a significant effect ($p < .01$), and there was also a significant ($p < .001$) block effect.

We have seen in the ANOVA table that on one level the analysis of variance consists of splitting up a sum of squares into various components. Occasionally the subject is approached from this viewpoint—as algebraic manipulation with sums of squares. The geometric point of view (i.e., projections and the Pythagorean theorem) explains why the decomposition arises. Here, however, we approach the subject from a sum of squares point of view. The SS terms are examples of quadratic forms. See Appendix 1 for a short review of quadratic forms.

DEFINITION 5.1 The rank of a quadratic form is the rank of the matrix of the quadratic form.

LEMMA 5.1 Let $\mathbf{Q}$ be the n by n matrix of a quadratic form. Since $\mathbf{Q}$ is assumed to be symmetric, it may be diagonalized by the Principal Axis theorem. Let $\mathbf{PQP}' = \mathbf{D}$, where

$$\mathbf{D} = \begin{pmatrix} d_1 & 0 & \cdots & 0 \\ 0 & d_2 & & 0 \\ \vdots & & \ddots & \vdots \\ 0 & 0 & \cdots & d_n \end{pmatrix}$$

is a diagonal matrix whose diagonal elements are the eigenvalues of $\mathbf{Q}$ and the rows of the orthogonal matrix $\mathbf{P}$ are the normalized eigenvectors of $\mathbf{Q}$. Let $\mathbf{P}' = (\mathbf{p}_1\mathbf{p}_2 \cdots \mathbf{p}_n)$. Then

(1) rank $\mathbf{Q}$ = number of nonzero $d_i \equiv k$;
(2) let $\mathbf{p}_{i_1}, \ldots, \mathbf{p}_{i_k}$ correspond to the nonzero d_i; then $\mathbf{X}'\mathbf{QX}$ is a function of $\mathbf{P}_V$, the projection onto V, the subspace with orthonormal basis $\mathbf{p}_{i_1}, \ldots, \mathbf{p}_{i_k}$;
(3) if all the $d_i = 0$ or 1, then $\mathbf{X}'\mathbf{QX} = \|\mathbf{P}_V\mathbf{X}\|^2$, where $\mathbf{P}_V$ is the projection onto V;
(4) if we can write the quadratic form $f(\mathbf{X})$ as a function of m linear forms $\boldsymbol{l}_1'\mathbf{X}, \ldots, \boldsymbol{l}_m'\mathbf{X}$ for any vectors $\boldsymbol{l}_1, \ldots, \boldsymbol{l}_m$, then rank $\mathbf{Q} \leq m$.

Proof (1) Recall that the rank of a matrix is unaffected by pre- or post-multiplication by a nonsingular matrix. Since $\mathbf{P}$ and $\mathbf{P}'$ are orthogonal and hence nonsingular, the result follows.

(2) Since we may write the matrix $\mathbf{Q}$ as $\sum_{i=1}^{n} d_i \mathbf{p}_i \mathbf{p}_i'$, then

$$\begin{aligned}\mathbf{X}'\mathbf{Q}\mathbf{X} &= \sum_{i=1}^{n} d_i(\mathbf{p}_i'\mathbf{X})^2 = d_{i_1}(\mathbf{p}_{i_1}'\mathbf{X})^2 + \cdots + d_{i_k}(\mathbf{p}_{i_k}'\mathbf{X})^2 \\ &= d_{i_1}((\mathbf{P}_V\mathbf{X})'\mathbf{p}_{i_1})^2 + \cdots + d_{i_k}((\mathbf{P}_V\mathbf{X})'\mathbf{p}_{i_k})^2\end{aligned}$$

since $(\mathbf{P}_V\mathbf{X})'\mathbf{p}_{i_1} = \mathbf{p}_{i_1}'\mathbf{X}, \ldots, (\mathbf{P}_V\mathbf{X})'\mathbf{p}_{i_k} = \mathbf{p}_{i_k}'\mathbf{X}$.

(3) If $d_{i_1} = d_{i_2} = \cdots = d_{i_k} = 1$ and other $d_i = 0$, then

$$d_{i_1}(\mathbf{p}_{i_1}'\mathbf{X})^2 + \cdots + d_{i_k}(\mathbf{p}_{i_k}'\mathbf{X})^2 = (\mathbf{p}_{i_1}'\mathbf{X})^2 + \cdots + (\mathbf{p}_{i_k}'\mathbf{X})^2 = \|\mathbf{P}_V\mathbf{X}\|^2$$

by Problem 3.4.

(4) The rank of a matrix does not change if the matrix is examined in a different coordinate system. Consider $\mathbf{Q}$ as represented in an orthonormal coordinate system with basis $\mathbf{v}_1, \ldots, \mathbf{v}_{m'}, \ldots, \mathbf{v}_n$, where $\mathbf{v}_1, \ldots, \mathbf{v}_{m'}$ span the subspace generated by $\boldsymbol{l}_1, \ldots, \boldsymbol{l}_m$ $(m' \le m)$. For any vector $\mathbf{X} = \sum_{i=1}^{n} \alpha_i \mathbf{v}_i$, $\mathbf{X}$ has the same value for the quadratic form as $\sum_{i=1}^{m'} \alpha_i \mathbf{v}_i$, since the inner product with the $\boldsymbol{l}_i$'s is the same. Thus, in this basis the quadratic form is

$$\sum_{i=1}^{m'} \sum_{j=1}^{m'} \alpha_i \alpha_j \tilde{q}_{ij},$$

and the transformed matrix is

$$\tilde{\mathbf{Q}} = \begin{array}{c} \\ m' \\ \\ \\ n-m' \\ \\ \end{array} \left(\begin{array}{ccc|ccc} \tilde{q}_{1,1} & \cdots & \tilde{q}_{1,m'} & 0 & \cdots & 0 \\ \vdots & & \vdots & \vdots & & \vdots \\ \tilde{q}_{m',1} & \cdots & \tilde{q}_{m',m'} & 0 & \cdots & 0 \\ \hline 0 & \cdots & 0 & 0 & \cdots & 0 \\ \vdots & & \vdots & \vdots & & \vdots \\ 0 & \cdots & 0 & 0 & \cdots & 0 \end{array}\right)$$

(column blocks: m' and $n - m'$)

so that rank $\le m' \le m$. □

Example Consider the identity $\sum_{i=1}^{n} X_i^2 = n\bar{X}^2 + \sum_{i=1}^{n}(X_i - \bar{X})^2$. The first term is

$$n\bar{X}^2 = \frac{1}{n}\left(\sum_{i=1}^{n} X_i\right)^2 = \frac{1}{n}\mathbf{X}'\mathbf{1}\,\mathbf{1}'\mathbf{X} = \mathbf{X}'\mathbf{Q}\mathbf{X},$$

where $\mathbf{Q} \equiv (1/n)\mathbf{1}\,\mathbf{1}'$ is the matrix of the quadratic form. Since $n\bar{X}^2 = (1/n)(\mathbf{1}'\mathbf{X})^2$, $n\bar{X}^2$ is a function of the linear form $\mathbf{1}'\mathbf{X}$ and thus $\mathbf{Q}$ has rank ≤ 1;

in fact, rank $\mathbf{Q} = 1$. (Why?) The second term is

$$\sum_{i=1}^{n} (X_i - \bar{X})^2 = \mathbf{X}'\left(\mathbf{I} - \frac{1}{n}\mathbf{1}\,\mathbf{1}'\right)\mathbf{X}.$$

Let $\boldsymbol{l}_i = \boldsymbol{\delta}_i - (1/n)\mathbf{1}$, so that $\boldsymbol{l}_i'\mathbf{X} = X_i - \bar{X}$. Since $\sum_{i=1}^{n} \boldsymbol{l}_i'\mathbf{X} = \sum_{i=1}^{n} (X_i - \bar{X}) = 0$, letting $\boldsymbol{l}_n'\mathbf{X} = -\sum_{i=1}^{n-1} \boldsymbol{l}_i'\mathbf{X}$, we can write

$$\sum_{i=1}^{n} (X_i - \bar{X})^2 = \sum_{i=1}^{n} (\boldsymbol{l}_i'\mathbf{X})^2 = \sum_{i=1}^{n-1} (\boldsymbol{l}_i'\mathbf{X})^2 + \left(-\sum_{i=1}^{n-1} \boldsymbol{l}_i'\mathbf{X}\right)^2,$$

so that $\sum_{i=1}^{n} (X_i - \bar{X})^2$ can be written as a function of the $n - 1$ linear forms $\boldsymbol{l}_1'\mathbf{X}, \ldots, \boldsymbol{l}_{n-1}'\mathbf{X}$, and thus rank $(\sum_{i=1}^{n} (X_i - \bar{X})^2) \leq n - 1$. Using the following results, we will see that these quadratic forms are the squares of the lengths of projections onto orthogonal subspaces.

THEOREM 5.1 Let V_1 and V_2 be subspaces of R^n with respective projections $\mathbf{P}_1$ and $\mathbf{P}_2$ and dimensions d_1 and d_2 such that $d_1 + d_2 \leq n$. If there exists real-valued functions f_1 and f_2 with domain R^n such that $f_1(\mathbf{0}) = f_2(\mathbf{0}) = 0$ and for all $\mathbf{x}$ in R^n, $\|\mathbf{x}\|^2 = f_1(\mathbf{P}_1\mathbf{x}) + f_2(\mathbf{P}_2\mathbf{x})$, then

(1) $d_1 + d_2 = n$,
(2) V_1 and V_2 are orthogonal subspaces,
(3) $f_1(\mathbf{P}_1\mathbf{x}) = \|\mathbf{P}_1\mathbf{x}\|^2$ and $f_2(\mathbf{P}_2\mathbf{x}) = \|\mathbf{P}_2\mathbf{x}\|^2$.

Proof (1) Suppose $d_1 + d_2 < n$. Then there is an $\mathbf{x} \neq \mathbf{0}$ such that $\mathbf{x} \perp V_1$ and $\mathbf{x} \perp V_2$ where $\perp$ denotes "perpendicular to." Then

$$\|\mathbf{x}\|^2 = f_1(\mathbf{P}_1\mathbf{x}) + f_2(\mathbf{P}_2\mathbf{x}) = f_1(\mathbf{0}) + f_2(\mathbf{0}) = 0,$$

a contradiction, so that $d_1 + d_2 = n$.

Let $W_2 = \{\mathbf{y} | \mathbf{y} \perp V_1\}$ (see Figure 5.1). W_2 is the orthogonal complement of V_1. $\dim W_2 = n - d_1 = d_2$. Let $\mathbf{w}$ be in W_2 and $\mathbf{P}_2\mathbf{w} = \mathbf{0}$; then

$$\|\mathbf{w}\|^2 = f_1(\mathbf{P}_1\mathbf{w}) + f_2(\mathbf{P}_2\mathbf{w}) = f_1(\mathbf{0}) + f_2(\mathbf{0}) = 0$$

so that $\mathbf{w} = \mathbf{0}$. Thus $\mathbf{P}_2$ restricted to a linear transformation from W_2 into V_2 maps only $\mathbf{0}$ into $\mathbf{0}$ and is a 1–1 onto transformation as the two subspaces have same dimension. Let $\mathbf{P}_2^{-1}$ denote the unique inverse transformation. Let $\mathbf{v}$ be in V_2 and $\mathbf{w} = \mathbf{P}_2^{-1}\mathbf{v}$. Then as $\mathbf{w}$ is perpendicular to V_1,

$$\|\mathbf{w}\|^2 = f_1(\mathbf{P}_1\mathbf{w}) + f_2(\mathbf{P}_2\mathbf{w}) = f_2(\mathbf{v}).$$

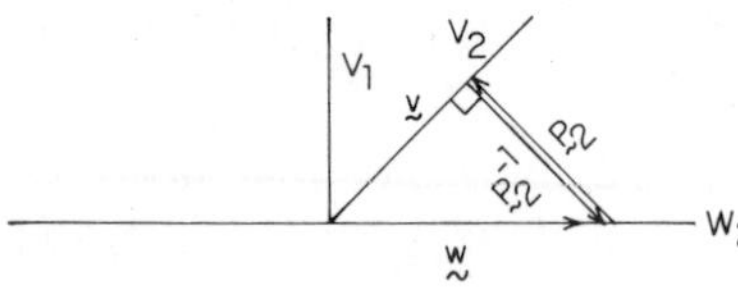

FIGURE 5.1

For arbitrary $\mathbf{v}$, $\mathbf{P}_2\mathbf{v}$ is in V_2, and letting $\mathbf{w} = \mathbf{P}_2^{-1}\mathbf{P}_2\mathbf{v}$, we have

$$f_2(\mathbf{P}_2\mathbf{v}) = \|\mathbf{P}_2^{-1}\mathbf{P}_2\mathbf{v}\|^2. \tag{5a}$$

Interchanging the indices 1 and 2, we obtain

$$f_1(\mathbf{P}_1\mathbf{v}) = \|\mathbf{P}_1^{-1}\mathbf{P}_1\mathbf{v}\|^2. \tag{5b}$$

Suppose $V_2 \neq W_2$. Then choose $\mathbf{v}$ in W_2, $\mathbf{v}$ not in V_2. Then

$$\begin{aligned}\|\mathbf{v}\|^2 > \|\mathbf{P}_2\mathbf{v}\|^2 &= f_1(\mathbf{P}_1\mathbf{P}_2\mathbf{v}) + f_2(\mathbf{P}_2\mathbf{P}_2\mathbf{v}) \geq f_2(\mathbf{P}_2\mathbf{v}) \\ &= \|\mathbf{P}_2^{-1}\mathbf{P}_2\mathbf{v}\|^2 = \|\mathbf{v}\|^2,\end{aligned}$$

where the second inequality follows from (5b) (showing that the f_1 term is ≥ 0) and $\mathbf{P}_2{}^2 = \mathbf{P}_2$ (as $\mathbf{P}_2$ is a projection). This contradiction shows that $V_2 = W_2$ and $\mathbf{P}_2^{-1}$ is the identity operation on V_2 and

$$f_2(\mathbf{P}_2\mathbf{v}) = \|\mathbf{P}_2^{-1}\mathbf{P}_2\mathbf{v}\|^2 = \|\mathbf{P}_2\mathbf{v}\|^2. \quad \square$$

Example In our one-sample t-test,

$$\sum_{i=1}^{n} X_i{}^2 = \underset{\substack{\uparrow \\ \text{rank} = 1}}{n\bar{X}^2} + \sum_{i=1}^{n} \underset{\substack{\uparrow \\ \text{rank} \leq n-1}}{(X_i - \bar{X})^2}.$$

Thus, the sums of squares are squared lengths of orthogonal projections and when divided by σ^2 are noncentral χ^2 random variables.

THEOREM 5.2 Let $V_1, \ldots, V_k$ be subspaces of R^n with respective projections $\mathbf{P}_1, \ldots, \mathbf{P}_k$ and dimensions $d_1, \ldots, d_k$ such that $d_1 + d_2 + \cdots + d_k \leq n$. If there exists k real-valued functions f_i: $R^n \to R$ such that $f_i(\mathbf{0}) = 0$ for $i = 1, \ldots, k$ and for all $\mathbf{x}$ in R^n,

$$\|\mathbf{x}\|^2 = f_1(\mathbf{P}_1\mathbf{x}) + f_2(\mathbf{P}_2\mathbf{x}) + \cdots + f_k(\mathbf{P}_k\mathbf{x}),$$

then

(1) $d_1 + d_2 + \cdots + d_k = n$,
(2) $V_1, \ldots, V_k$ are mutually orthogonal subspaces,
(3) $f_i(\mathbf{P}_i\mathbf{x}) = \|\mathbf{P}_i\mathbf{x}\|^2$.

Proof Problem 5.12. $\square$

THEOREM 5.3 (*Cochran's theorem*) Let $\mathbf{X} \sim N(\boldsymbol{\mu}, \sigma^2\mathbf{I})$, and if for all $\mathbf{X}$ in R^n, $\|\mathbf{X}\|^2 = \mathbf{X}'\mathbf{Q}_1\mathbf{X} + \cdots + \mathbf{X}'\mathbf{Q}_k\mathbf{X}$, where the $\mathbf{X}'\mathbf{Q}_i\mathbf{X}$ are quadratic forms of rank d_i and $d_1 + \cdots + d_k \leq n$, then

(1) the quadratic forms are the squared norms of projections onto orthogonal subspaces whose dimensions add to n, that is, $\mathbf{X}'\mathbf{Q}_i\mathbf{X} = \|\mathbf{P}_i\mathbf{X}\|^2$, where $\mathbf{P}_i$ is the projection onto the subspace V_i of dimension d_i and $V_1, \ldots, V_k$ are mutually orthogonal subspaces whose dimensions add to n.

(2) the k quadratic forms are independent noncentral chi-squared random variables with d_i df and noncentrality parameters $\delta_i = \sqrt{\boldsymbol{\mu}'\mathbf{Q}_i\boldsymbol{\mu}}/\sigma$.

Proof Problem 5.13. □

Example (*Two-sample t statistic*) Suppose we observe $X_1, \ldots, X_n$ i.i.d. $N(\mu_1, \sigma^2)$ and independently $Y_1, \ldots, Y_m$ i.i.d. $N(\mu_2, \sigma^2)$. Further suppose that μ_1, μ_2, and σ^2 are unknown and that we want to test $\mu_1 = \mu_2$ versus $\mu_1 \neq \mu_2$.

Consider the $(n+m)$-dimensional vector $\mathbf{X}' = (X_1, \ldots, X_n, Y_1, \ldots, Y_m)$. Then

$$\|\mathbf{X}\|^2 = \sum_{i=1}^{n} X_i^2 + \sum_{j=1}^{m} Y_j^2 = n\bar{X}^2 + m\bar{Y}^2 + \sum_{i=1}^{n}(X_i - \bar{X})^2 + \sum_{j=1}^{m}(Y_j - \bar{Y})^2$$

$$= \underset{\text{rank } 1}{\frac{nm}{n+m}(\bar{X} - \bar{Y})^2} + \underset{\text{rank } 1}{\frac{1}{n+m}(n\bar{X} + m\bar{Y})^2} + \underset{\text{rank } n-1}{\sum_{i=1}^{n}(X_i - \bar{X})^2} + \underset{\text{rank } m-1}{\sum_{j=1}^{m}(Y_j - \bar{Y})^2}.$$

To show the first two terms have rank 1 use

$$\boldsymbol{l}_1' = \left(\frac{1}{n}, \ldots, \frac{1}{n}, -\frac{1}{m}, \ldots, -\frac{1}{m}\right), \qquad \boldsymbol{l}_2' = (1, 1, \ldots, 1).$$

If $\mu_1 = \mu_2$, the noncentrality parameter squared for the first quadratic form is

$$\frac{nm}{n+m}\,\frac{(\mu_1 - \mu_2)^2}{\sigma^2} = 0.$$

Thus, use

$$F = \frac{[nm/(n+m)](\bar{X} - \bar{Y})^2/1}{\left(\sum_{i=1}^{n}(X_i - \bar{X})^2 + \sum_{j=1}^{m}(Y_j - \bar{Y})^2\right)/(n+m-2)}$$

to test $\mu_1 = \mu_2$.

Example (*Balanced one-way analysis of variance*) Suppose we have I treatments and take J observations on each treatment. Suppose the jth measurement on the ith treatment is $X_{ij} \sim N(\mu_i, \sigma^2), i = 1, \ldots, I; j = 1, \ldots, J$, and these random variables are independent. Let $\mu \equiv (1/I)\sum_{i=1}^{I} \mu_i$:

$$\sum_{i=1}^{I}\sum_{j=1}^{J} X_{ij}^2 = \sum_i\sum_j (X_{ij} - \bar{X}_{i\cdot} + \bar{X}_{i\cdot} - \bar{X}_{\cdot\cdot} + \bar{X}_{\cdot\cdot})^2$$

$$= \sum_i\sum_j (X_{ij} - \bar{X}_{i\cdot})^2 + J\sum_i(\bar{X}_i - \bar{X}_{\cdot\cdot})^2 + IJ\bar{X}_{\cdot\cdot}^2,$$

where the quadratic forms are $\mathbf{Q}_1$, $\mathbf{Q}_2$, and $\mathbf{Q}_3$ respectively.

TABLE 5.4

Source	df	SS	MS	E(MS)
Treatments	$I-1$	$SS_T = J\sum_i(\bar{X}_{i\cdot} - \bar{X}_{\cdot\cdot})^2$	$SS_T/(I-1)$	$\sigma^2 + J\sum_i(\mu_i - \mu)^2/(I-1)$
Error	$I(J-1)$	$SS_e = \sum_i\sum_j(X_{ij} - \bar{X}_{i\cdot})^2$	$SS_e/I(J-1)$	σ^2
Grand mean	1	$SS_{GM} = IJ\bar{X}_{\cdot\cdot}^2$	$SS_{GM}/1$	
Total	IJ	$\sum_i\sum_j X_{ij}^2$		

TABLE 5.5

Source	df	SS	MS	E(MS)
Treatments	$I-1$	SS_T	$SS_T/(I-1)$	$\sigma^2 + J\sum_i(\mu_i - \mu)^2/(I-1)$
Error	$I(J-1)$	SS_e	$SS_e/I(J-1)$	σ^2
Total	$IJ-1$	$\sum_i\sum_j(X_{ij} - \bar{X}_{\cdot\cdot})^2$		

Rank $\mathbf{Q}_1 \leq I(J-1)$, Rank $\mathbf{Q}_2 \leq I-1$, Rank $\mathbf{Q}_3 \leq 1$, and as $I(J-1) + (I-1) + 1 = IJ$, Cochran's theorem applies. The ANOVA table is given in Table 5.4. Often the grand mean is omitted so that the ANOVA table becomes that shown in Table 5.5.

SUMMARY

The geometric point of view (i.e., projections and the Pythagorean theorem) explains why the analysis of variance consists of splitting up a sum of squares into various components. Occasionally the subject is approached from a sum of squares point of view, that is, as an algebraic manipulation with sum of squares. The terms are examples of quadratic forms.

THEOREM Let $\mathbf{Q}$ be the matrix of a quadratic form which without loss of generality may be taken to be symmetric. Then

(1) the rank of $\mathbf{Q}$ equals the number of nonzero eigenvalues of $\mathbf{Q}$;

(2) the quadratic form is a function of the projection onto a subspace whose dimension equals the rank of the quadratic form;

(3) if all the eigenvalues are zero or one, then $\mathbf{X}'\mathbf{Q}\mathbf{X} = \|\mathbf{P}\mathbf{X}\|^2$, where $\mathbf{P}$ is the projection onto the subspace spanned by the eigenvectors corresponding to the nonzero eigenvalues of $\mathbf{Q}$;

(4) if we can write the quadratic form as a function of m linear forms, then the rank of $\mathbf{Q}$ is less than or equal to m.

COCHRAN'S THEOREM Let $\mathbf{X} \sim N(\boldsymbol{\mu}, \sigma^2\mathbf{I})$, and if for all $\mathbf{X}$ in R^n, $\|\mathbf{X}\|^2 = \mathbf{X}'\mathbf{Q}_1\mathbf{X} + \cdots + \mathbf{X}'\mathbf{Q}_k\mathbf{X}$, where the $\mathbf{X}'\mathbf{Q}_i\mathbf{X}$ are quadratic forms of rank d_i and $d_1 + \cdots + d_k \leq n$, then

(a) the quadratic forms are the squared norms of projections onto orthogonal subspaces whose dimensions add to n; that is, $\mathbf{X}'\mathbf{Q}_i\mathbf{X} = \|\mathbf{P}_i\mathbf{X}\|^2$, where $\mathbf{P}_i$ is a projection on a subspace V_i of dimension d_i; $V_1, \ldots, V_k$ are mutually orthogonal subspaces whose dimensions add to n;
(b) the k quadratic forms divided by σ^2, $\mathbf{X}'\mathbf{Q}_i\mathbf{X}/\sigma^2$, $i = 1, \ldots, k$, are independent noncentral chi-squared random variables with d_i df and noncentrality parameters $\delta_i = \sqrt{\boldsymbol{\mu}'\mathbf{Q}_i\boldsymbol{\mu}}/\sigma$.

PROBLEMS

5.1 Show that the model of equation (1) is unique; that is, given terms $c_{ij} = \mu + \alpha_i + \beta_j$, there are no other $\tilde{\mu} + \tilde{\alpha}_i + \tilde{\beta}_j$.

5.2 Show that $S = \{$set of vectors $\mathbf{v}$ such that $\mathbf{v} = \alpha_1\mathbf{v}_1 + \cdots + \alpha_I\mathbf{v}_I$ for some $\alpha_1, \ldots, \alpha_I$, where $\sum_{i=1}^{I} \alpha_i = 0\}$ is an $(I-1)$-dimensional subspace of R^{IJK}. (*Note*: $\mathbf{v}_i \notin S$.)

5.3 Show that the subspace of Problem 5.2 has as a basis $\{\mathbf{v}_1 - \mathbf{v}_I, \ldots, \mathbf{v}_{I-1} - \mathbf{v}_I\}$. Note that these vectors are *not* orthogonal.

5.4 Show that the three subspaces mentioned in the comments following equation (2) are mutually orthogonal.

5.5 Consider a subspace S with basis $\mathbf{b}_1, \ldots, \mathbf{b}_m$. Show that the orthogonal projection of a vector $\mathbf{v}$ onto S is the unique vector in S which has the same inner product as $\mathbf{v}$ with each of $\mathbf{b}_1, \ldots, \mathbf{b}_m$.

5.6 The text uses the notation

$$\begin{pmatrix} \mathbf{u}_1' \\ \vdots \\ \mathbf{u}_{I-1}' \end{pmatrix} (\mathbf{u}_1, \ldots, \mathbf{u}_{I-1}).$$

For $I = 3$, $J = 2$, and $K = 2$, write out the two matrices explicitly. Multiply them and verify this special case of equation (3). Check also that the special case of the inverse as given following equation (3) is correct in this case.

5.7 Verify equation (4).

5.8 In explaining the ANOVA table, the statement is made that the mean square term is the sum of squares term "adjusted" for dimension. In what sense is this true?

5.9 Justify the distributional assertion for testing the hypothesis $\alpha_i \equiv 0$ and $\beta_j \equiv 0$; that is, the test statistic $(\mathrm{SS}_A + \mathrm{SS}_B)/\mathrm{MS}_e(I + J - 2)$ has the given distribution.

5.10 (a) Note that the square of a $t_{n-1}(\delta)$ random variable is a $N(\delta, 1)^2/\chi^2_{n-1}/(n-1)$ random variable and the two random variables, that is, N, χ^2, are independent (see Problem 3.18).

(b) Use (a) to construct appropriate ANOVA tables for the two-sample t statistics of earlier chapters.

5.11 If $f(x_1, x_2) = (x_1, x_2)\binom{3\ 1}{1\ 3}\binom{x_1}{x_2}$, find an orthonormal basis that diagonalizes the quadratic form.

5.12 Prove Theorem 5.2. [*Hint*: If $V_1' = V_1$ and V_2' is the vector space generated by $V_2 \cup \cdots \cup V_k$, show that V_1' and V_2' satisfy Theorem 5.1. In particular, to show $f_2'(\mathbf{P}_2{}^*\mathbf{x}) = f_2(\mathbf{P}_2\mathbf{x}) + \cdots + f_k(\mathbf{P}_k\mathbf{x})$, show that $\mathbf{P}_j\mathbf{x} = \mathbf{P}_j\mathbf{P}_2{}^*\mathbf{x}$, where $\mathbf{P}_2{}^*$ is the projection onto V_2'.

5.13 Prove Theorem 5.3.

5.14 (*Differing methods of choosing parameters*) In this chapter it was necessary to put constraints on the α_i and β_j ($\sum_i \alpha_i = \sum_j \beta_j = 0$) in order to give an identifiable model. There are an infinite number of ways to choose such constraints to give identifiability; what if another system of constraints had been chosen?

Much confusion can be avoided in this area if we keep the geometric interpretation before us. It is best to think of each hypothesis as specifying that the mean vector of the observations lies in a specified subspace.

Consider our model

$$E(X_{ij}) = \mu_{ij} = \mu + \alpha_i + \beta_j, \qquad \sum_i \alpha_i = \sum_j \beta_j = 0.$$

Decide whether each of the following pairs of restriction in place of $\sum_i \alpha_i = 0$, $\sum_j \beta_j = 0$ leads to the "same" model; that is, the same subspace contains the mean vector:

(a) $\alpha_1 = 0, \beta_1 = 0$,
(b) $\alpha_1 = 1, \beta_1 = 1$,
(c) $\alpha_1 = \alpha_2 = 0$.

5.15 For the part(s) of Problem 5.14 which give the same model, relate the parameters under the original restriction and the new restriction given in the problem.

Consider the hypotheses restricting the mean (expressed in terms of the model of the text) $H_A: \alpha_i = 0$ for all i and $H_B: \beta_j = 0$ for all j. For the alternative parametrizations given above, what are these hypotheses framed in terms of the alternative method of parametrizing the problem?

5.16* In many analyses of variance situations it is useful to make more than one measurement on the same experimental unit. In such a situation one speaks of a *repeated measures* design, experiment, or analysis. To illustrate the point, consider the following example. In a two-way ANOVA, if there is considerable expense in preparing experimental units and the measurement has a large measurement error but may be made quickly and cheaply, economy would dictate making multiple measurements on each unit to reduce the effect of measurement error. (Note that there is still random error above and beyond the measurement error because of the inherent variability between experimental units.)

Assuming a nice balanced situation and independence between measurement error and the remaining variability, how would you model and analyze such data?

5.17 (*Repeated measures continued*) Another *repeated measures* situation arises when different treatments are tried on the same individuals. In a biomedical situation, this may arise for several reasons: (1) the availability of individuals needing or wanting to participate may be limited; (2) by trying the different treatments on the same individuals, comparisons may be made taking out the variability between individuals (see the model below); (3) the logistics of involving additional individuals may be quite complex.

In using different treatments on the same individuals, there should be no "carry-over" effects between treatments or this must be taken into account in the analysis and the analysis given below would be inappropriate. Suppose we have I individuals each receiving J treatments with a response X_{ij} being measured. A model which is often appropriate is

$$X_{ij} = \mu + \alpha_i + \beta_j + e_{ij},$$

where $i = 1, \ldots, I, j = 1, \ldots, J, \sum_j \beta_j = 0$, and the α_i and e_{ij} are $I + IJ$ mutually independent normal random variables with mean zero; $\alpha_i \sim N(0, \sigma_A{}^2)$, $e_{ij} \sim N(0, \sigma^2)$.

(a) Show that the ANOVA table, Table 5.6, is appropriate (including independence of the sums of squares).

(b) How do you test for no treatment effect (i.e., $\beta_j \equiv 0$)?

(c) Suppose different individuals are selected for each treatment. Find the expected mean square for the appropriate one-way analysis of variance.

(d) If the model of this problem is correct, under what conditions is the repeated measures design more efficient?

TABLE 5.6

SS	df	E(MS)
SS_A	$I - 1$	$\sigma^2 + J\sigma_A{}^2$
SS_B	$J - 1$	$\sigma^2 + I\sum_j \beta_j{}^2/(J - 1)$
SS_e	$(I - 1)(J - 1)$	σ^2

Example [P. P. Engler and J. A. Bowers, Vitamin B_6 in reheated, held, and freshly cooked turkey breast, *Journal of the American Dietetic Association* **67**, No. 1 (July 1975)] As an example utilizing the analysis of this problem, consider the work of Engler and Bowers, in which they were able to implement a repeated measures design for their study. Since in many food service establishments cooked meats are often held in warming ovens and on steamtables or are refrigerated and reheated before serving, the authors investigated whether the holding and/or reheating process may result in loss of vitamin B_6 content in turkey breast quarters.

Sixteen frozen breast halves from eight 22- to 24-lb Grade A tom turkeys from the same lot and processed under similar conditions were obtained from a local plant. The two breast halves from one turkey were divided while frozen and the resulting four quarters were thawed and assigned to the four treatments: (a) roasted; (b) roasted, refrigerated 24 hr, then reheated in an electric oven; (c) roasted, refrigerated 24 hr, then reheated in a microwave oven; (d) roasted, sliced, and held in a warming oven for 60 min and then in a warming tray for 15 min. At the end of each treatment, samples to be analyzed for vitamin B_6 were freeze-dried.

The design of Problem 5.17 with each turkey having four repeated measurements was the experimental design. Vitamin B_6 data were analyzed by analysis of variance. That analysis removed turkey variation from the desired comparisons. When vitamin B_6 was calculated on the basis of cooked weight, no significant differences were observed among treatments; however, when calculated on a moisture-free and fat-free basis, significant differences were observed among treatments. For both measures, variation among birds was greater than variation among treatments. The authors did not present the complete ANOVA table but only the mean vitamin B_6 levels for each treatment and the significance of the F values for treatment and bird (block) sources of variation (see Table 5.7).

TABLE 5.7

Mean Vitamin B_6 Content of Turkey Breast Muscle

					Significance of F value	
	Roasted	Roasted, reheated, electric oven	Roasted, reheated, microwave oven	Roasted, held	Treatment	Blocks (birds)
Vitamin B_6 (mcg/g muscle)	4.30	3.69	3.79	3.61	Not significant	$p < .001$
Vitamin B_6 (mcg/g muscle, moisture-free, fat-free basis)	15.57	12.12	11.78	11.57	$p < .05$	$p < .001$

5.18 (*Robustness continued*) In the case in which $k = 1$ (one observation per cell), Friedman has introduced a rank test of the null hypothesis that all $\beta_j = 0$. For each of the I rows, rank the J observations. Let r_{ij} be the rank of the i,j observation within its own row. *Friedman's chi-square statistic* is

$$X^2 = \frac{12}{IJ(J+1)}\left[\sum_{j=1}^{J}\left(\sum_{i=1}^{I} r_{ij}\right)^2\right] - 3I(J+1).$$

As $I \to \infty$, under the null hypothesis $\beta_j \equiv 0$, X^2 has a limiting $\chi^2_{J-1}(0)$-distribution. Compute X^2 for the data of Problem 5.19.

5.19 (*Example 5.2 continued*) The plasma-free cystine level in rats fed heated soybean protein (μmoles/100 ml) is presented in Table 5.8 (where each observation is the average for four rats). Calculate the ANOVA table for this new response variable.

TABLE 5.8

	Heat treatment					
Group	1	2	3	4	5	6
I (block I)	4.0	4.0	4.1	3.8	4.5	3.8
II (block II)	4.6	5.7	5.2	4.9	5.6	5.3
III (block III)	4.9	6.1	5.4	5.2	5.9	5.7

6 ESTIMATION AND MORE ON FACTORIAL DESIGNS

We have seen that in many situations, for example, testing whether a number of treatments give the same mean effect, it is possible to test various hypotheses using appropriate F-tests derived from the ANOVA. In most situations, however, it is not testing per se that is desired but rather to estimate the various effects involved. For example, with a number of treatments we might wish to estimate the effect of each treatment—not to test that they all have the same effect. Or, if we have rejected the hypothesis that all have the same effect, we might want to know which treatment had the greatest effect. Estimation has been thinly veiled beneath the surface of our tests.

How in general do we construct good estimates? Intuitive estimates of parameters relating to the mean vector $E(\mathbf{X})$ are those values of the parameters which result in a mean vector nearest the observation vector. Are these intuitive estimates any good? How, in general, do we construct good estimates? How accurate are the estimates? What if we want to compare many different estimates?

Note that if $\mathbf{X}$ is multinormally distributed, we may write our models as $\mathbf{X} = \mathbf{D}\boldsymbol{\beta} + \mathbf{e}$, where $\mathbf{e} \sim N(\mathbf{0}, \sigma^2\mathbf{I})$. Note that this model is linear in the unknown parameters $\boldsymbol{\beta}$.

Examples For the two-sample t statistic of Chapter 3,

$$\mathbf{X}' = (X_1, \ldots, X_n, Y_1, \ldots, Y_m),$$

$$\mathbf{D} = \begin{matrix} n \text{ positions} \left\{ \vphantom{\begin{matrix}1\\ \vdots\\ 1\end{matrix}} \right. \\ m \text{ positions} \left\{ \vphantom{\begin{matrix}0\\ \vdots\\ 0\end{matrix}} \right. \end{matrix} \begin{pmatrix} 1 & 0 \\ \vdots & \vdots \\ 1 & 0 \\ 0 & 1 \\ \vdots & \vdots \\ 0 & 1 \end{pmatrix}, \quad \text{and} \quad \boldsymbol{\beta} = \begin{pmatrix} \mu \\ \omega \end{pmatrix}.$$

For the one-way ANOVA of Chapter 4,

$$\mathbf{X}' = (X_{11}, \ldots, X_{1n_1}, X_{21}, \ldots, X_{2n_2}, \ldots, X_{k1}, \ldots, X_{kn_k}),$$

$$\mathbf{D} = \begin{matrix} n_1\left\{\begin{matrix}\\ \\ \\ \end{matrix}\right. \\ n_2\left\{\begin{matrix}\\ \\ \\ \end{matrix}\right. \\ \vdots \\ n_k\left\{\begin{matrix}\\ \\ \\ \end{matrix}\right. \end{matrix} \begin{pmatrix} 1 & 0 & 0 & \cdots & 0 \\ \vdots & \vdots & & & \vdots \\ 1 & 0 & 0 & \cdots & 0 \\ 0 & 1 & 0 & \cdots & 0 \\ \vdots & \vdots & \vdots & & \vdots \\ 0 & 1 & 0 & \cdots & 0 \\ \vdots & \vdots & \vdots & & \vdots \\ 0 & 0 & 0 & \cdots & 1 \\ \vdots & \vdots & \vdots & & \vdots \\ 0 & 0 & 0 & \cdots & 1 \end{pmatrix}, \quad \text{and} \quad \boldsymbol{\beta} = \begin{pmatrix} \mu_1 \\ \mu_2 \\ \vdots \\ \mu_k \end{pmatrix}.$$

Our models then are written as $E(\mathbf{X}) = \mathbf{D}\boldsymbol{\beta}$, where $\mathbf{D}$ is a n by p matrix, called the *design matrix*, and $\boldsymbol{\beta}$ is a p by 1 vector of unknown parameters. The covariance of $\mathbf{X}$ is assumed to be $\sigma^2\mathbf{I}$. Clearly $\mathbf{D}\boldsymbol{\beta}$ is a vector in the subspace spanned by the column vectors $\mathbf{d}_1, \ldots, \mathbf{d}_p$ of $\mathbf{D}$ as $\mathbf{D}\boldsymbol{\beta} = (\mathbf{d}_1, \ldots, \mathbf{d}_p)\boldsymbol{\beta} = \beta_1\mathbf{d}_1 + \cdots + \beta_p\mathbf{d}_p$. The $\boldsymbol{\beta}$'s cannot be determined uniquely if the $\mathbf{d}_i$'s are linearly dependent. Thus we consider the case in which the $\mathbf{d}_i$'s are linearly independent or equivalently the design matrix has full column rank p, the dimension of $\boldsymbol{\beta}$. The least squares estimates of $\beta_1, \ldots, \beta_p$, the parameters relating to the mean vector $\mathbf{D}\boldsymbol{\beta}$, are those values of the parameters, denoted $\hat{\beta}_1, \ldots, \hat{\beta}_p$, which result in a mean vector nearest the observation vector. Since the mean vector lies in the estimation space M which equals the subspace spanned by the columns of the design matrix, the least squares estimate $\hat{\boldsymbol{\beta}}$ of $\boldsymbol{\beta}$ must satisfy $\mathbf{D}\hat{\boldsymbol{\beta}} = \mathbf{P}_M\mathbf{X}$, where $\mathbf{P}_M$ is the projection onto the estimation space M. We seek to find the form of $\mathbf{P}_M$.

THEOREM 6.1 If M is the subspace spanned by the linearly independent column vectors of the matrix $\mathbf{D}$, then the projection onto M is

$$\mathbf{P}_M = \mathbf{D}(\mathbf{D}'\mathbf{D})^{-1}\mathbf{D}'.$$

Proof $\mathbf{D}'\mathbf{D}$ has an inverse since $\mathbf{D}'\mathbf{D}\mathbf{z} = \mathbf{0}$ implies $\mathbf{z}'\mathbf{D}'\mathbf{D}\mathbf{z} = \mathbf{0}$. Thus, $\|\mathbf{D}\mathbf{z}\|^2 = 0$ and $\mathbf{D}\mathbf{z} = \mathbf{0}$. As the columns of $\mathbf{D}$ are linearly independent and $\mathbf{D}\mathbf{z} = z_1\mathbf{d}_1 + \cdots + z_p\mathbf{d}_p = \mathbf{0}$, this implies that all $z_i = 0$. Thus, the $\mathbf{z}'\mathbf{D}'\mathbf{D}\mathbf{z} \geq 0$ and equal to zero only if $\mathbf{z} = \mathbf{0}$; that is, $\mathbf{D}'\mathbf{D}$ is a positive definite matrix and has an inverse (Theorem A34). To finish the proof note that $\mathbf{y} = \mathbf{P}_M\mathbf{y} + (\mathbf{I} - \mathbf{P}_M)\mathbf{y}$. Clearly, $\mathbf{P}_M\mathbf{y}$ is in M. We need to check that $(\mathbf{I} - \mathbf{P}_M)\mathbf{y}$ is orthogonal to every column of $\mathbf{D}$ or $\mathbf{0} = \mathbf{D}'(\mathbf{I} - \mathbf{P}_M)\mathbf{y}$, but

$$\mathbf{D}'(\mathbf{I} - \mathbf{P}_M)\mathbf{y} = \mathbf{D}'(\mathbf{I} - \mathbf{D}(\mathbf{D}'\mathbf{D})^{-1}\mathbf{D}')\mathbf{y} = (\mathbf{D}' - \mathbf{D}')\mathbf{y} = \mathbf{0},$$

so we are done. Note that $\mathbf{D}(\mathbf{D}'\mathbf{D})^{-1}\mathbf{D}'$ is a symmetric idempotent matrix. □

The least squares estimate $\hat{\boldsymbol{\beta}}$ of $\boldsymbol{\beta}$ satisfies $\mathbf{D}\hat{\boldsymbol{\beta}} = \mathbf{P}_M\mathbf{X} = \mathbf{D}(\mathbf{D}'\mathbf{D})^{-1}\mathbf{D}'\mathbf{X}$, so that $\hat{\boldsymbol{\beta}} = (\mathbf{D}'\mathbf{D})^{-1}\mathbf{D}'\mathbf{X}$ is the least squares estimate of $\boldsymbol{\beta}$.

THEOREM 6.2 If $E(\mathbf{X}) = \mathbf{D}\boldsymbol{\beta}$, where $\mathbf{D}$ has full column rank p, the dimension of $\boldsymbol{\beta}$, and $\text{cov}(\mathbf{X}) = \sigma^2\mathbf{I}$, then

(1) the least squares estimator of $\boldsymbol{\beta}$, $\hat{\boldsymbol{\beta}} = (\mathbf{D}'\mathbf{D})^{-1}\mathbf{D}'\mathbf{X}$, is an unbiased estimator of $\boldsymbol{\beta}$, that is, $E(\hat{\boldsymbol{\beta}}) = \boldsymbol{\beta}$;
(2) $\text{cov}(\hat{\boldsymbol{\beta}}) = \sigma^2(\mathbf{D}'\mathbf{D})^{-1}$;
(3) $\min_{\tilde{\boldsymbol{\beta}}}\|\mathbf{X} - \mathbf{D}\tilde{\boldsymbol{\beta}}\|^2 = \|\mathbf{X} - \mathbf{D}\hat{\boldsymbol{\beta}}\|^2$;
(4) *GAUSS–MARKOV THEOREM* The least squares estimator has "minimum variance" within the class of linear unbiased estimators of $\boldsymbol{\beta}$.
Let $\tilde{\boldsymbol{\beta}} = \mathbf{T}\mathbf{X}$ be a linear unbiased estimator of $\boldsymbol{\beta}$. Then

(i) let $\text{cov}(\tilde{\boldsymbol{\beta}})$ and $\text{cov}(\hat{\boldsymbol{\beta}})$ be the two covariance matrices of the estimators $\tilde{\boldsymbol{\beta}}$ and $\hat{\boldsymbol{\beta}}$. Then $\text{cov}(\tilde{\boldsymbol{\beta}}) - \text{cov}(\hat{\boldsymbol{\beta}})$ is nonnegative definite (this says that the dispersion of the random vector $\hat{\boldsymbol{\beta}}$ about its mean vector $\boldsymbol{\beta}$ is smaller than that of
(ii) $\text{var}(\mathbf{c}'\tilde{\boldsymbol{\beta}}) \geq \text{var}(\mathbf{c}'\hat{\boldsymbol{\beta}})$ for all p by 1 vectors $\mathbf{c}$. (This says that the variance of any linear combination of the components of $\tilde{\boldsymbol{\beta}}$ is larger than that of the same linear combination of the components of $\hat{\boldsymbol{\beta}}$; hence choosing $\mathbf{c} = \boldsymbol{\delta}_i$ for $i = 1, \ldots, p$, we have $\text{var}(\tilde{\beta}_i) \geq \text{var}(\hat{\beta}_i)$.)

Proof (1) $E(\hat{\boldsymbol{\beta}}) = E((\mathbf{D}'\mathbf{D})^{-1}\mathbf{D}'\mathbf{X}) = (\mathbf{D}'\mathbf{D})^{-1}\mathbf{D}'E(\mathbf{X}) = (\mathbf{D}'\mathbf{D})^{-1}\mathbf{D}'\mathbf{D}\boldsymbol{\beta} = \boldsymbol{\beta}$.
(2) $\text{cov}(\hat{\boldsymbol{\beta}}) = \text{cov}((\mathbf{D}'\mathbf{D})^{-1}\mathbf{D}'\mathbf{X}) = (\mathbf{D}'\mathbf{D})^{-1}\mathbf{D}'(\sigma^2\mathbf{I})[(\mathbf{D}'\mathbf{D})^{-1}\mathbf{D}']' = \sigma^2(\mathbf{D}'\mathbf{D})^{-1}$.
(3) $\|\mathbf{X} - \mathbf{D}\tilde{\boldsymbol{\beta}}\|^2 = \|\mathbf{X} - \mathbf{D}\hat{\boldsymbol{\beta}} + \mathbf{D}\hat{\boldsymbol{\beta}} - \mathbf{D}\tilde{\boldsymbol{\beta}}\|^2 = \|(\mathbf{X} - \mathbf{D}\hat{\boldsymbol{\beta}}) + \mathbf{D}(\hat{\boldsymbol{\beta}} - \tilde{\boldsymbol{\beta}})\|^2 = \|\mathbf{X} - \mathbf{D}\hat{\boldsymbol{\beta}}\|^2 + \|\mathbf{D}(\hat{\boldsymbol{\beta}} - \tilde{\boldsymbol{\beta}})\|^2$ since $\mathbf{X} - \mathbf{D}\hat{\boldsymbol{\beta}}$ is orthogonal to the estimation space M and $\mathbf{D}(\hat{\boldsymbol{\beta}} - \tilde{\boldsymbol{\beta}})$ is in M.
(4) Part (i): We need to show that for each p by 1 vector $\mathbf{c}$, $\mathbf{c}'(\text{cov}(\tilde{\boldsymbol{\beta}}) - \text{cov}(\hat{\boldsymbol{\beta}}))\mathbf{c} \geq 0$ or $\mathbf{c}'\,\text{cov}(\tilde{\boldsymbol{\beta}})\mathbf{c} \geq \mathbf{c}'\,\text{cov}(\hat{\boldsymbol{\beta}})\mathbf{c}$. But since $\text{var}(\mathbf{c}'\tilde{\boldsymbol{\beta}}) = \mathbf{c}'\,\text{cov}(\tilde{\boldsymbol{\beta}})\mathbf{c}$ and $\text{var}(\mathbf{c}'\hat{\boldsymbol{\beta}}) = \mathbf{c}'\,\text{cov}(\hat{\boldsymbol{\beta}})\mathbf{c}$, we note that part (ii) is equivalent to part (i). To show part (ii), since each row of the p by n matrix $\mathbf{T}$ can be written as a sum of two orthogonal vectors, one in the estimation space M plus one in the error space $\mathscr{E}$, we can write $\mathbf{T} = \mathbf{T}_M + \mathbf{T}_{\mathscr{E}}$, where the rows of $\mathbf{T}_M$ are in the estimation space M and the rows of $\mathbf{T}_{\mathscr{E}}$ are orthogonal to the estimation space. Now

$$E(\tilde{\boldsymbol{\beta}}) = E(\mathbf{T}\mathbf{X}) = \mathbf{T}E(\mathbf{X}) = \mathbf{T}_M\mathbf{D}\boldsymbol{\beta} + \mathbf{T}_{\mathscr{E}}\mathbf{D}\boldsymbol{\beta} = \mathbf{T}_M\mathbf{D}\boldsymbol{\beta} + \mathbf{0} = \mathbf{T}_M\mathbf{D}\boldsymbol{\beta} = \boldsymbol{\beta}$$

for all $\boldsymbol{\beta}$ since $\tilde{\boldsymbol{\beta}}$ is unbiased. Since $(\mathbf{D}'\mathbf{D})^{-1}\mathbf{D}'$ and $\mathbf{T}_M$ have rows in the estimation space, the rows of $\mathbf{T}_M - (\mathbf{D}'\mathbf{D})^{-1}\mathbf{D}'$ are also in the estimation space. But they are also in the error space since $E(\mathbf{T}_M\mathbf{X}) = \boldsymbol{\beta}$, and

$E((\mathbf{D}'\mathbf{D})^{-1}\mathbf{D}'\mathbf{X}) = \boldsymbol{\beta}$ implies for all $\mathbf{D}\boldsymbol{\beta}$,

$$\mathbf{0} = \boldsymbol{\beta} - \boldsymbol{\beta} = E([\mathbf{T}_M\mathbf{X} - (\mathbf{D}'\mathbf{D})^{-1}\mathbf{D}'\mathbf{X}]) = [\mathbf{T}_M - (\mathbf{D}'\mathbf{D})^{-1}\mathbf{D}']E(\mathbf{X}).$$

Since the only vector in two orthogonal subspaces is the $\mathbf{0}$ vector, we have $\mathbf{T}_M = (\mathbf{D}'\mathbf{D})^{-1}\mathbf{D}'$. Therefore

$$\begin{aligned}\operatorname{var}(\mathbf{c}'\tilde{\boldsymbol{\beta}}) &= \operatorname{var}(\mathbf{c}'\mathbf{T}\mathbf{X}) = \operatorname{var}(\mathbf{c}'[\mathbf{T}_M + \mathbf{T}_{\mathscr{E}}]\mathbf{X}) = \operatorname{var}(\mathbf{c}'\mathbf{T}_M\mathbf{X} + \mathbf{c}'\mathbf{T}_{\mathscr{E}}\mathbf{X}) \\ &= \operatorname{var}(\mathbf{c}'\mathbf{T}_M\mathbf{X}) + \operatorname{var}(\mathbf{c}'\mathbf{T}_{\mathscr{E}}\mathbf{X}) + 2\operatorname{cov}(\mathbf{c}'\mathbf{T}_M\mathbf{X}, \mathbf{c}'\mathbf{T}_{\mathscr{E}}\mathbf{X}) \\ &= \operatorname{var}(\mathbf{c}'\hat{\boldsymbol{\beta}}) + \operatorname{var}(\mathbf{c}'\mathbf{T}_{\mathscr{E}}\mathbf{X}) \geq \operatorname{var}(\mathbf{c}'\hat{\boldsymbol{\beta}})\end{aligned}$$

since $\operatorname{cov}(\mathbf{c}'\mathbf{T}_M\mathbf{X}, \mathbf{c}'\mathbf{T}_{\mathscr{E}}\mathbf{X}) = \mathbf{c}'\mathbf{T}_M(\sigma^2\mathbf{I})\mathbf{T}_{\mathscr{E}}'\mathbf{c} = \sigma^2\mathbf{c}'\mathbf{T}_M\mathbf{T}_{\mathscr{E}}'\mathbf{c} = 0$ since $\mathbf{T}_M\mathbf{T}_{\mathscr{E}}' = \mathbf{0}$.

COROLLARY Let $\mathbf{D} = (\mathbf{d}_1, \ldots, \mathbf{d}_p)$ have orthogonal columns. Then the least squares estimates $\hat{\boldsymbol{\beta}}_1, \ldots, \hat{\boldsymbol{\beta}}_p$ have the form $\hat{\boldsymbol{\beta}}_i = \mathbf{d}_i'\mathbf{X}/\|\mathbf{d}_i\|^2$ and all are independent if $\mathbf{X} \sim N(\mathbf{D}\boldsymbol{\beta}, \sigma^2\mathbf{I})$.

Proof

$$\mathbf{D}'\mathbf{D} = \begin{pmatrix} \|\mathbf{d}_1\|^2 & 0 & \cdots & 0 \\ 0 & \|\mathbf{d}_2\|^2 & \cdots & 0 \\ \vdots & \vdots & & \vdots \\ 0 & 0 & \cdots & \|\mathbf{d}_p\|^2 \end{pmatrix},$$

$$(\mathbf{D}'\mathbf{D})^{-1} = \begin{pmatrix} \frac{1}{\|\mathbf{d}_1\|^2} & 0 & \cdots & 0 \\ 0 & \frac{1}{\|\mathbf{d}_2\|^2} & \cdots & 0 \\ \vdots & \vdots & & \vdots \\ 0 & 0 & \cdots & \frac{1}{\|\mathbf{d}_p\|^2} \end{pmatrix},$$

$$(\mathbf{D}'\mathbf{D})^{-1}\mathbf{D}'\mathbf{X} = \begin{pmatrix} \frac{1}{\|\mathbf{d}_1\|^2} & 0 & \cdots & 0 \\ 0 & \frac{1}{\|\mathbf{d}_2\|^2} & \cdots & 0 \\ \vdots & \vdots & & \vdots \\ 0 & 0 & \cdots & \frac{1}{\|\mathbf{d}_p\|^2} \end{pmatrix} \begin{pmatrix} \mathbf{d}_1'\mathbf{X} \\ \mathbf{d}_2'\mathbf{X} \\ \vdots \\ \mathbf{d}_p'\mathbf{X} \end{pmatrix} = \begin{pmatrix} \frac{\mathbf{d}_1'\mathbf{X}}{\|\mathbf{d}_1\|^2} \\ \frac{\mathbf{d}_2'\mathbf{X}}{\|\mathbf{d}_2\|^2} \\ \vdots \\ \frac{\mathbf{d}_p'\mathbf{X}}{\|\mathbf{d}_p\|^2} \end{pmatrix} = \hat{\boldsymbol{\beta}}.$$

The $\hat{\boldsymbol{\beta}}_i$ are independent as they are functions of projections onto orthogonal subspaces. □

Note that none of these results including the Gauss–Markov theorem required the assumption that $\mathbf{X}$ was multinormally distributed!

THEOREM 6.3 If $\mathbf{X} \sim N(\mathbf{D}\boldsymbol{\beta}, \sigma^2\mathbf{I})$, where $\mathbf{D}$ has full column rank, then

(i) the least squares estimator $\hat{\boldsymbol{\beta}}$ is the maximum likelihood estimator of $\boldsymbol{\beta}$, and
(ii) $\hat{\boldsymbol{\beta}} \sim N(\boldsymbol{\beta}, \sigma^2(\mathbf{D}'\mathbf{D})^{-1})$.

Proof (i) The likelihood function is

$$L(\mathbf{X}) = |\sigma^2\mathbf{I}|^{-1/2}(2\pi)^{-n/2}\exp\left(-\frac{1}{2\sigma^2}\|\mathbf{X}-\mathbf{D}\boldsymbol{\beta}\|^2\right).$$

To maximize this, we minimize $\|\mathbf{X}-\mathbf{D}\boldsymbol{\beta}\|^2$, so (i) follows from Theorem 6.2, part (3), and for (ii) do Problem 6.1. □

Example (*Two-way ANOVA of Chapter 5*) $X_{ij} = \mu + \alpha_i + \beta_j + e_{ij}$. To get a full rank $\mathbf{D}$, we replace α_I by $-\sum_{i=1}^{I-1}\alpha_i$ and β_J by $-\sum_{j=1}^{J-1}\beta_j$. The column of $\mathbf{D}$ for μ has IJ 1's. The column of $\mathbf{D}$ for α_i has 0's except that

(i) the J terms of $X_{ij}, j = 1, \ldots, J$, have a 1;
(ii) the J terms of $X_{Ij}, j = 1, \ldots, J$, have a -1.

The column of $\mathbf{D}$ for β_j has 0's except that

(i) the I terms of $X_{ij}, i = 1, \ldots, I$, have a 1;
(ii) the I terms of $X_{iJ}, i = 1, \ldots, I$, have a -1.

For example, if $I = 3$, $J = 2$, then

$$\mathbf{X} = \begin{pmatrix} X_{11} \\ X_{12} \\ X_{21} \\ X_{22} \\ X_{31} \\ X_{32} \end{pmatrix}, \qquad \boldsymbol{\beta} = \begin{pmatrix} \mu \\ \alpha_1 \\ \alpha_2 \\ \beta_1 \end{pmatrix}, \qquad \mathbf{D} = \begin{pmatrix} 1 & 1 & 0 & 1 \\ 1 & 1 & 0 & -1 \\ 1 & 0 & 1 & 1 \\ 1 & 0 & 1 & -1 \\ 1 & -1 & -1 & 1 \\ 1 & -1 & -1 & -1 \end{pmatrix},$$

$$\mathbf{D}'\mathbf{D} = \begin{pmatrix} 1 & 1 & 1 & 1 & 1 & 1 \\ 1 & 1 & 0 & 0 & -1 & -1 \\ 0 & 0 & 1 & 1 & -1 & -1 \\ 1 & -1 & 1 & -1 & 1 & -1 \end{pmatrix} \begin{pmatrix} 1 & 1 & 0 & 1 \\ 1 & 1 & 0 & -1 \\ 1 & 0 & 1 & 1 \\ 1 & 0 & 1 & -1 \\ 1 & -1 & -1 & 1 \\ 1 & -1 & -1 & -1 \end{pmatrix} = \begin{pmatrix} 6 & 0 & 0 & 0 \\ 0 & 4 & 2 & 0 \\ 0 & 2 & 4 & 0 \\ 0 & 0 & 0 & 6 \end{pmatrix},$$

$$(\mathbf{D}'\mathbf{D})^{-1} = \begin{pmatrix} \frac{1}{6} & 0 & 0 & 0 \\ 0 & \frac{1}{3} & -\frac{1}{6} & 0 \\ 0 & -\frac{1}{6} & \frac{1}{3} & 0 \\ 0 & 0 & 0 & \frac{1}{6} \end{pmatrix},$$

$$\mathbf{D}' = \begin{pmatrix} 1 & 1 & 1 & 1 & 1 & 1 \\ 1 & 1 & 0 & 0 & -1 & -1 \\ 0 & 0 & 1 & 1 & -1 & -1 \\ 1 & -1 & 1 & -1 & 1 & -1 \end{pmatrix}, \quad \mathbf{D}'\mathbf{X} = \begin{pmatrix} 6\bar{X}_{..} \\ 2(\bar{X}_{1.} - \bar{X}_{3.}) \\ 2(\bar{X}_{2.} - \bar{X}_{3.}) \\ 3(\bar{X}_{.1} - \bar{X}_{.2}) \end{pmatrix},$$

$$(\mathbf{D}'\mathbf{D})^{-1}\mathbf{D}'\mathbf{X} = \begin{pmatrix} \frac{1}{6} & 0 & 0 & 0 \\ 0 & \frac{1}{3} & -\frac{1}{6} & 0 \\ 0 & -\frac{1}{6} & \frac{1}{3} & 0 \\ 0 & 0 & 0 & \frac{1}{6} \end{pmatrix} \begin{pmatrix} 6\bar{X}_{..} \\ 2(\bar{X}_{1.} - \bar{X}_{3.}) \\ 2(\bar{X}_{2.} - \bar{X}_{3.}) \\ 3(\bar{X}_{.1} - \bar{X}_{.2}) \end{pmatrix}$$

$$= \begin{pmatrix} \bar{X}_{..} \\ \frac{2}{3}\bar{X}_{1.} - \frac{1}{3}(\bar{X}_{3.} + \bar{X}_{2.}) \\ \frac{2}{3}\bar{X}_{2.} - \frac{1}{3}(\bar{X}_{1.} + \bar{X}_{3.}) \\ \frac{1}{2}(\bar{X}_{.1} - \bar{X}_{.2}) \end{pmatrix} = \begin{pmatrix} \bar{X}_{..} \\ \bar{X}_{1.} - \bar{X}_{..} \\ \bar{X}_{2.} - \bar{X}_{..} \\ \bar{X}_{.1} - \bar{X}_{..} \end{pmatrix}.$$

Note $\bar{X}_{3.} + \bar{X}_{2.} = 3\bar{X}_{..} - \bar{X}_{1.}$, etc.

Analyses of variance usually result from planned experiments. The matrix **D** is known as it is "chosen or designed" by the experimenter.

The matrix algebra is usually done by computer. For balanced designs the form of the answers is known and need not be recomputed each time (e.g., $\bar{X}_{i.} - \bar{X}_{..}$). Let us return to the two-factorial designs of the previous chapter and consider another possible use for them. To do this we shall want to consider the idea of *randomization* in experimentation. Randomization in an experiment occurs when some procedure or treatment or allocation of units in the experiment is done by using some random device. For example, in an experiment comparing two different drugs, half the people will often receive one drug and half another drug. The half that receives each drug is chosen at random from among the totality of subjects. The idea of randomization, due to Sir R. A. Fisher, is one of the great intellectual advances in the methodology of science. Randomization has many important virtues.

(1) It guards statistically against important *unknown* factors, so that all treatments that are considered see "comparable" units *on the average*. The way it guards against the unknown factors is to allow one to statistically take account of the probability that a given effect might be due to such factors.

(2) It protects against human biases subtle and not so subtle in experimentation.

(3) The probabilistic theory underlying randomization allows estimation of the effect of errors upon the evaluation of the experiment.

Often in randomizing, however, one does not, even when comparing several treatments select the groups at random. If there are factors that are known to affect the outcome, the experimental units are often put into blocks, with the blocks having similar values of the factor known to affect the outcome, and then the randomization is done within the blocks. This allows more efficient estimation of various parameters, a topic we turn to in Chapter 9.

Consider giving three drugs and a control to mice and recording the amount of sleep (assumed normally distributed). We might give each of the four possibilities to five animals chosen at random from the 20 animals, and then do a one-way analysis of variance. Suppose on second thought, however, we feel that weight should be an important variable and will contribute quite heavily to the variability. To remove much of this variability, we might arrange the four heaviest mice in one block, the next four heaviest in another block, etc. In each block, we assign the treatments at random to the four mice, so that each treatment is given to one of the four mice. We might model this *randomized block* design by the model of Chapter 5, $X_{ij} = \mu + \alpha_i + \beta_j + e_{ij}$. The usual assumptions $\sum_i \alpha_i = 0$ and $\sum_j \beta_j = 0$ are made as well as the assumption that the e_{ij}'s are i.i.d. $N(0, \sigma^2)$ random variables.

To think of what this additive model means, consider graphing the means (Figures 6.1 and 6.2).

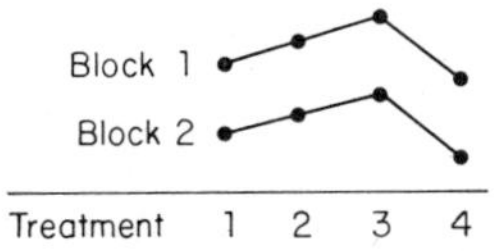

FIGURE 6.1 For an equal distance apart for each treatment, the additive model holds.

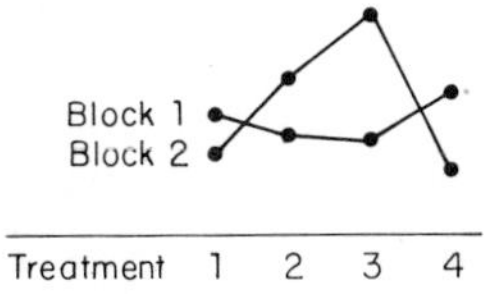

FIGURE 6.2 The additive model does not hold (there is an interaction between blocks and treatments).

Suppose now, however, that the additive model does not seem to hold. What can we do? For this we turn to models allowing an interaction.

Example (*Complete balanced two-way factorial design*) Let factor A have I levels ($i = 1, \ldots, I$), factor B have J levels ($j = 1, \ldots, J$), and let there be K replications ($k = 1, \ldots, K$) at each possible combination of the levels of the factors. Let X_{ijk} be independent $N(\mu_{ij}, \sigma^2)$, $i = 1, \ldots, I$, $j = 1, \ldots, J$, $k = 1, \ldots, K$, or $X_{ijk} = \mu_{ij} + e_{ijk}$, e_{ijk} i.i.d. $N(0, \sigma^2)$. We can write

$$X_{ijk} = \mu + \alpha_i + \beta_j + \gamma_{ij} + e_{ijk}, \qquad \text{where } \sum_i \alpha_i = \sum_j \beta_j = \sum_i \gamma_{ij} = \sum_j \gamma_{ij} = 0,$$

$$\mu \equiv \bar{\mu}_{\cdot\cdot}, \qquad \alpha_i \equiv \bar{\mu}_{i\cdot} - \bar{\mu}_{\cdot\cdot}, \qquad \beta_j \equiv \bar{\mu}_{\cdot j} - \bar{\mu}_{\cdot\cdot}, \qquad \gamma_{ij} \equiv \mu_{ij} - \bar{\mu}_{i\cdot} - \bar{\mu}_{\cdot j} + \bar{\mu}_{\cdot\cdot}.$$

DEFINITION 6.1 A factorial design is *complete* if each possible combination of levels of the factors has at least one observation. Note that the preceding design is complete. To construct the ANOVA table when $K > 1$,

$$\sum_i\sum_j\sum_k X_{ijk}^2 = \sum_{i,j,k} (\underbrace{X_{ijk} - \bar{X}_{ij\cdot}}_{\text{error}} + \underbrace{\bar{X}_{ij\cdot} - \bar{X}_{i\cdot\cdot} - \bar{X}_{\cdot j\cdot} + \bar{X}_{\cdot\cdot\cdot}}_{\text{interaction}} + \underbrace{\bar{X}_{i\cdot\cdot} - \bar{X}_{\cdot\cdot\cdot}}_{\alpha_i}$$

$$+ \underbrace{\bar{X}_{\cdot j\cdot} - \bar{X}_{\cdot\cdot\cdot}}_{\beta_j} + \underbrace{\bar{X}_{\cdot\cdot\cdot}}_{\text{grand mean}})^2$$

$$= \underbrace{\sum_{i,j,k} (X_{ijk} - \bar{X}_{ij\cdot})^2}_{\text{SS}_\text{e}} + \underbrace{K \sum_{ij} (\bar{X}_{ij\cdot} - \bar{X}_{i\cdot\cdot} - \bar{X}_{\cdot j\cdot} + \bar{X}_{\cdot\cdot})^2}_{\text{SS}_\text{int}}$$

$$+ \underbrace{JK \sum_i (\bar{X}_{i\cdot\cdot} - \bar{X}_{\cdot\cdot\cdot})^2}_{\text{SS}_A} + \underbrace{IK \sum_j (\bar{X}_{\cdot j\cdot} - \bar{X}_{\cdot\cdot\cdot})^2}_{\text{SS}_B} + \underbrace{IJK\bar{X}^2_{\cdot\cdot\cdot}}_{\text{SS}_{GM}} \quad (1)$$

The ANOVA table is given in Table 6.1 (note that the grand mean is omitted). To test

$$H_0\colon\ \gamma_{ij} = 0, \qquad \text{use } \text{MS}_\text{int}/\text{MS}_\text{e} \sim F_{(I-1)(J-1), IJ(K-1)};$$
$$H_0\colon\ \alpha_i = 0, \qquad \text{use } \text{MS}_A/\text{MS}_\text{e} \sim F_{I-1, IJ(K-1)}.$$

TABLE 6.1 $I \times J$ Factorial Design

Source	SS	df	MS	E(MS)
A	SS_A	$I-1$	$\text{SS}_A/(I-1)$	$\sigma^2 + JK\sum_i \alpha_i^2/(I-1)$
B	SS_B	$J-1$	$\text{SS}_B/(J-1)$	$\sigma^2 + IK\sum_j \beta_j^2/(J-1)$
Interaction (AB)	SS_int	$(I-1)(J-1)$	$\text{SS}_\text{int}/(I-1)(J-1)$	$\sigma^2 + K\sum_{i,j} \gamma_{ij}^2/(I-1)(J-1)$
Error	SS_e	$IJ(K-1)$	$\text{SS}_\text{e}/IJ(K-1)$	σ^2
Total	$\sum_i\sum_j\sum_k (X_{ijk} - \bar{X}_{\cdot\cdot\cdot})^2$	$IJK-1$		

If the interaction effect is large, the interpretation of the main effects must be considered very carefully. In practice, the interaction terms are often found to be zero or negligibly small.

Why would one often expect the interaction effects to be negligible in a two-factor design? Suppose the levels of factor A occur by choosing specific values of a quantity x and the levels of B arise by specification of a quantity y (e.g., in a chemical engineering experiment, x might be temperature and y might be the concentration of a chemical). For each pair of values (x, y) we measure a response, say X. Suppose $E(X) = f(x, y)$, where the *response surface* $f(x, y)$ is a "smooth" function of x and y.

Suppose the experiment consists of looking at I values for x, say $x_1, \ldots, x_I$ (the levels of factor A), and J values for y, say $y_1, \ldots, y_J$ (the levels of factor B). Suppose further that the experiment has K repetitions of the IJ combinations. A reasonable model is

$$X_{ijk} = f(x_i, y_j) + e_{ijk},$$
$$i = 1, 2, \ldots, I, \qquad j = 1, \ldots, J, \qquad k = 1, \ldots, K;$$
$$e_{ijk} \quad \text{i.i.d.} \quad N(0, \sigma^2).$$

As f is a smooth function, if the x_i and y_j do not vary greatly about $\bar{x} \equiv (1/I)\sum_i x_i$, $\bar{y} \equiv (1/J)\sum_j y_j$, Taylor's theorem in two variables tells us that f may be reasonably approximated about $\bar{x}$ and $\bar{y}$ by a linear function; that is, there exist numbers a and b such that

$$f(x_i, y_j) \approx f(\bar{x}, \bar{y}) + a(x_i - \bar{x}) + b(y_j - \bar{y})$$

where $a = \partial f(\bar{x}, \bar{y})/\partial x$ and $b = \partial f(\bar{x}, \bar{y})/\partial y$. Thus, our model is (approximately)

$$X_{ijk} = \mu + \alpha_i + \beta_j + e_{ijk},$$

where $\mu = f(\bar{x}, \bar{y})$, $\alpha_i = a(x_i - \bar{x})$, and $\beta_j = b(y_j - \bar{y})$. In summary, if the "response surface" f is smooth, approximation of the surface by a plane leads to an analysis of variance with no interaction term.

As noted previously, it is an empirical fact that in a large fraction (but by no means all) of the ANOVA's run, the interaction terms are not statistically significant.

Example We move on to a complete balanced three-way factorial design Let factor A have I levels ($i = 1, \ldots, I$), factor B have J levels ($j = 1, \ldots, J$), factor C have K levels ($k = 1, \ldots, K$), and let there be L replications ($l = 1, \ldots, L$) at each possible combination of the levels of the factors:

$$X_{ijkl} = \mu_{ijk} + e_{ijkl},$$
$$i = 1, \ldots, I, \qquad j = 1, \ldots, J, \qquad k = 1, \ldots, K, \qquad l = 1, \ldots, L;$$
$$e_{ijkl} \quad \text{i.i.d.} \quad N(0, \sigma^2).$$

As before, we break μ_{ijk} up into a grand mean, main effects, two-factor interactions, and the three-factor interaction which must be defined to give any possible μ_{ijk} pattern:

$$\mu \equiv \bar{\mu}_{...}, \qquad \alpha_i \equiv \bar{\mu}_{i..} - \bar{\mu}_{...}, \qquad \beta_j \equiv \bar{\mu}_{.j.} - \bar{\mu}_{...}, \qquad \gamma_k \equiv \bar{\mu}_{..k} - \bar{\mu}_{...},$$
$$(\alpha\beta)_{ij} \equiv \bar{\mu}_{ij.} - \bar{\mu}_{i..} - \bar{\mu}_{.j.} + \bar{\mu}_{...}, \qquad (\alpha\gamma)_{ik} \equiv \bar{\mu}_{i.k} - \bar{\mu}_{i..} - \bar{\mu}_{..k} + \bar{\mu}_{...},$$
$$(\beta\gamma)_{jk} \equiv \bar{\mu}_{.jk} - \bar{\mu}_{.j.} - \bar{\mu}_{..k} + \bar{\mu}_{...},$$

where we denote the AB two-factor interaction term by $(\alpha\beta)_{ij}$, etc. Let $(\alpha\beta\gamma)_{ijk}$ be the three-factor interaction. We want

$$\mu_{ijk} = \mu + \alpha_i + \beta_j + \gamma_k + (\alpha\beta)_{ij} + (\alpha\gamma)_{ik} + (\beta\gamma)_{jk} + (\alpha\beta\gamma)_{ijk}.$$

This implies that

$$(\alpha\beta\gamma)_{ijk} \equiv \mu_{ijk} - \bar{\mu}_{ij.} - \bar{\mu}_{i.k} - \bar{\mu}_{.jk} + \bar{\mu}_{i..} + \bar{\mu}_{.j.} + \bar{\mu}_{..k} - \bar{\mu}_{...}.$$

Another way of thinking of the three-factor interaction is that it is the difference between the two-factor AB interaction for the specific level k minus the two-factor interaction AB averaged over k (our two-factor interaction).

Thus,

$$(\alpha\beta\gamma)_{ijk} = (\mu_{ijk} - \bar{\mu}_{i.k} - \bar{\mu}_{.jk} + \bar{\mu}_{..k}) - (\mu_{ij.} - \bar{\mu}_{i..} - \bar{\mu}_{.j.} + \bar{\mu}_{...}).$$

"Crunching" through the SS, we find that

$$\sum_i \sum_j \sum_k \sum_l X_{ijkl}^2 = IJKL\bar{X}_{....}^2 + \mathrm{SS}_A + \mathrm{SS}_B + \mathrm{SS}_C$$
$$+ \mathrm{SS}_{AB} + \mathrm{SS}_{AC} + \mathrm{SS}_{BC} + \mathrm{SS}_{ABC} + \mathrm{SS}_{\mathrm{e}}.$$

The SS's come from the definitions of the effects; for example, as $(\alpha\beta)_{ij} = \bar{\mu}_{ij.} - \bar{\mu}_{i..} - \bar{\mu}_{.j.} + \bar{\mu}_{...}$, the corresponding SS is

$$\sum_i \sum_j \sum_k \sum_l \widehat{(\alpha\beta)}_{ij}^2 = KL \sum_i \sum_j (\bar{X}_{ij..} - \bar{X}_{i...} - \bar{X}_{.j..} + \bar{X}_{....})^2, \qquad \text{etc.}$$

Further, the terms in the SS's are the least squares estimates of the parameters involved. The ANOVA table that results is given in Table 6.2. Tests of appropriate hypotheses are clear; for example, test $\mu_{ijk} = \mu + \alpha_i + \beta_j + \gamma_k$. Use

$$F = \frac{(\mathrm{SS}_{AB} + \mathrm{SS}_{BC} + \mathrm{SS}_{AC} + \mathrm{SS}_{ABC})/\sum \mathrm{df}}{\mathrm{MS}_{\mathrm{e}}},$$

TABLE 6.2

ANOVA Table—$I \times J \times K$ Factorial Design

Source	SS	df	$E(\text{MS})$
A	SS_A	$I-1$	$\sigma^2 + JKL\sum_i \alpha_i^2/(I-1)$
B	SS_B	$J-1$	$\sigma^2 + IKL\sum_j \beta_j^2/(J-1)$
C	SS_C	$K-1$	$\sigma^2 + IJL\sum_k \gamma_k^2/(K-1)$
AB interaction	SS_{AB}	$(I-1)(J-1)$	$\sigma^2 + KL\sum_{i,j}(\alpha\beta)_{ij}^2/(I-1)(J-1)$
BC interaction	SS_{BC}	$(J-1)(K-1)$	$\sigma^2 + IL\sum_{j,k}(\beta\gamma)_{jk}^2/(J-1)(K-1)$
AC interaction	SS_{AC}	$(I-1)(K-1)$	$\sigma^2 + JL\sum_{i,k}(\alpha\gamma)_{ik}^2/(I-1)(K-1)$
ABC interaction	SS_{ABC}	$(I-1)(J-1)(K-1)$	$\sigma^2 + L\sum_{i,j,k}(\alpha\beta\gamma)_{ijk}^2/(I-1)(J-1)(K-1)$
Error	SS_e	$IJK(L-1)$	σ^2 where $SS_e = \sum_{i,j,k,l}(X_{ijkl} - \bar{X}_{ijk\cdot})^2$
Total	$\sum_{i,j,k,l}(X_{ijkl} - \bar{X}_{\cdots\cdot})^2$	$IJKL-1,$	

where

$$\begin{aligned}\sum \text{df} &= (I-1)(J-1) + (J-1)(K-1) + (I-1)(K-1) + (I-1)(J-1)(K-1)\\ &= IJK - I - J - K + 2.\end{aligned}$$

So F is $F_{IJK-I-J-K+2,IJK(L-1)}$.

Now turn to two examples of factorial designs.

Example 6.1 [B. Barrios, L. Corbitt, J. Estes, and J. Topping, Effect of a social stigma on interpersonal distance, *The Psychological Record* **26**, 343–348 (1976)] A large body of literature suggests that normal individuals tend to avoid interpersonal involvement with the physically stigmatized, that is, with those individuals who have physical characteristics with an intense negative appraisal. Unfortunately, relatively little attention has been devoted to the experimental analysis of the effects of social stigmas, that is, nonphysical, attributed behavioral characteristics which have a severe negative evaluation on interaction behavior.

Barrios *et al.* investigated the "free" selection of seating distances when normal subjects were involved in an informed interaction with either a socially stigmatized or nonstigmatized confederate. The social stigma investigated was bisexuality. It was hypothesized that, given a free seating choice, subjects interacting with a socially stigmatized confederate would select greater seating distances than would subjects interacting with a nonstigmatized confederate. Since there appears to be a tendency for females to prefer closer interaction distances than males, it was also predicted that females would select closer interaction distances than males, regardless of the experimental condition and sex of the confederate. Since previous research indicated that both female and male subjects prefer closer interaction distances with females, it was also expected that subjects would position themselves closer to the female confederate than the male confederate, regardless of the experimental condition and sex of the subject.

The experimental design was a $2 \times 2 \times 2$ factorial design in which the eight groups were defined by the sex of subject, sex of confederate, and experimental condition (stigmatized versus nonstigmatized). There were five subjects in each group. Prior to the subject's arrival, the experimenter randomly selected the experimental condition. The dependent variable was the subject's seating distance from the confederate during an interview situation. Mean group seating distances are presented in Table 6.3. Results of the $2 \times 2 \times 2$ analysis of variance revealed that subjects sat closer to a nonstigmatized than a stigmatized confederate and that subjects positioned themselves closer to a female confederate than to a male confederate. These outcomes were indicated by the significant main effects for experimental condition $F(1, 32) = 4.75$, $p < .05$, and sex of the confederate, $F(1, 32) = 9.13$, $p < .01$. The other main effect and all interaction effects were nonsignificant.

TABLE 6.3

Mean Group Seating Distances (in cm)

	Stigmatized		Nonstigmatized	
	Male	Female	Male	Female
Subjects	Confederate		Confederate	
Males	261.62	146.81	195.07	85.85
Females	326.14	193.55	206.25	124.97

Example 6.2 [S. Worchel, T. Hardy, and R. Hurley, The effects of commercial interruption of violent and nonviolent films on viewers' subsequent aggression, *Journal of Experimental Social Psychology* **12**, 220–232 (1976)] Considerable research attention has been given in recent years to the effects on subsequent aggressive behavior of viewing aggressive films. This research indicates that viewing violent films tends to increase subjects' aggressive behavior. Many previous studies of film-influenced aggression have obtained results in highly artificial situations, for example, short-duration film segments which portray nothing but violence. Studies in which entire films are

presented would provide a more sound base from which to generalize to actual viewing situations. In actual conditions, especially with television programming, the film may be interrupted by commercials. If commercial interruptions "break the spell of the viewer," the commercial interruption should decrease the aggression following an aggressive film. However, the commercials may be viewed as a source of frustration since they block the "ongoing drive" of the viewer to see the movie and thus may have an aggression-heightening effect.

Worchel *et al.* studied film-influenced aggression in a somewhat more coherent, realistic context. A balanced 3 × 2 factorial design was employed. Groups of subjects saw one of three full-length films: (1) a staged violent film with "acted" aggression, (2) a realistic violent film with scenes of actual violence, or (3) a nonviolent film. Each film was either interrupted by commercials or it was not.

All subjects were greeted by the experimenter who presented the experiment as a pilot study designed to collect data on a variety of topics relevant to motion pictures. Subjects were told that their role in the study would be to watch a movie and then to fill out a questionnaire. The experimenter introduced his experimental assistant explaining that he would tell the subjects more about the film, operate the projection equipment and distribute the questionnaires, and then the experimenter left the room. In the performance of his assigned tasks, the assistant committed three "blunders" that were designed to provide the subjects with some basis for criticizing his performance. After the subjects had viewed the film and completed the questionnaires, the assistant left the room; the experimenter returned, thanked the subjects for their participation, and elicited their help in one final matter. The experimenter explained that only one of his three assistants could be rehired next year and he wished to hire the most able person and that subject ratings were the best indicator of assistant performance. The experimenter asked the subjects to rate the assistant on a number of dimensions using a bipolar scale. The dimensions briefly were: (1) should hire (1 = strongly agree, . . . , 31 = strongly disagree), (2) overall effectiveness (1 = very effective, . . . , 31 = very ineffective), and (3) clarity of explanations (1 = very clear and concise, . . . , 31 = not at all clear and concise). It was felt that these ratings could be taken as a measure of aggression, not merely hostility, as subjects were informed that poor assistant ratings would keep him from getting the job. After completing the ratings, the subjects were fully debriefed.

Because groups of subjects were randomly assigned to experimental conditions and because each group was asked to rate the experimental assistant's behavior, which may have varied slightly from group to group, group average ratings were used as the unit of analysis. Subjects were run

TABLE 6.4

Means on Ratings of Assistant

	Wild one		Attica		Mouse that roared	
No commercial[a]	Should hire	9.54	Should hire	9.17	Should hire	7.38
	Effectiveness	10.40	Effectiveness	9.12	Effectiveness	7.21
	Clarity	10.47	Clarity	9.80	Clarity	8.22
Commercial	Should hire	16.63	Should hire	15.50	Should hire	8.79
	Effectiveness	15.86	Effectiveness	15.22	Effectiveness	9.39
	Clarity	15.30	Clarity	14.59	Clarity	8.93

[a] N = four per cell.

TABLE 6.5

Summary of Analyses

Source	df	Should hire F	Effectiveness F	Clarity F
Film (A)	2	22.88[b]	26.58[b]	21.29[b]
Commercial (B)	1	57.65[b]	64.06[b]	35.51[b]
AB	2	7.44[a]	4.46[a]	5.57[a]
Error (MS_e)	18	(2.55)	(1.97)	(2.01)

[a] $p < .05$
[b] $p < .001$.

in groups ranging in size from three to nine ($\bar{X} = 5.25$) persons and 126 subjects were used. There were four groups per cell. The assistant rating scale means are presented in Table 6.4 and the ANOVA tables are presented in Table 6.5. These results will be discussed further in Chapter 8.

SUMMARY

Our basic model is $E(\mathbf{X}) = \mathbf{D}\boldsymbol{\beta}$, where the design matrix $\mathbf{D}$ has full column rank p, the dimension of $\boldsymbol{\beta}$, our vector of unknown parameters. The covariance of $\mathbf{X}$ is assumed to be $\sigma^2\mathbf{I}$. $E(\mathbf{X})$ lies in the subspace spanned by the columns of the design matrix, called the estimation space. The projection operator onto the estimation space M is $\mathbf{P}_M = \mathbf{D}(\mathbf{D}'\mathbf{D})^{-1}\mathbf{D}'$. The parameters $\hat{\boldsymbol{\beta}}$ that satisfy $\mathbf{P}_M\mathbf{X} = \mathbf{D}\hat{\boldsymbol{\beta}}$ are called the least squares estimates of $\boldsymbol{\beta}$. Note that the least squares estimator $[\hat{\boldsymbol{\beta}} = (\mathbf{D}'\mathbf{D})^{-1}\mathbf{D}'\mathbf{X}]$ of $\boldsymbol{\beta}$ is an unbiased estimator of $\boldsymbol{\beta}$ with $\text{cov}(\hat{\boldsymbol{\beta}}) = \sigma^2(\mathbf{D}'\mathbf{D})^{-1}$. The least squares estimator has "minimum variance" within the class of linear unbiased estimators of $\boldsymbol{\beta}$

(Gauss–Markov theorem). None of the preceding results (including the Gauss–Markov theorem) requires the assumption that **X** is multinormally distributed.

The assumption that **X** is multinormally distributed is often made for the purpose of testing various hypotheses about our parameter vector $\boldsymbol{\beta}$.

THEOREM If $\mathbf{X} \sim N(\mathbf{D}\boldsymbol{\beta}, \sigma^2\mathbf{I})$, where **D** has full column rank, then

(i) the least squares estimator $\hat{\boldsymbol{\beta}}$ is the maximum likelihood estimator of $\boldsymbol{\beta}$, and

(ii) the least squares estimator is multinormally distributed, namely, $\hat{\boldsymbol{\beta}} \sim N(\boldsymbol{\beta}, \sigma^2(\mathbf{D}'\mathbf{D})^{-1})$.

PROBLEMS

6.1 Prove Theorem 6.3, part (ii).

6.2 For the two-factor ANOVA with interaction ($I = 2, J = 2, K = 2$), find the least squares estimates by using differential calculus methods.

6.3 In equation (1), verify that the cross-product terms drop out.

6.4 Check the df column for the ANOVA table for the two-factor design (Table 6.1) with interaction and use Cochran's theorem to verify that the table is correct.

6.5 Repeat Problem 6.4 for the three-factor ANOVA table (Table 6.2).

6.6 (*Simple linear regression using the least squares approach*) Consider the model

$$Y_i = \alpha + \beta x_i + e_i, \qquad e_i \quad \text{i.i.d.} \quad N(0, \sigma^2).$$

If this is observed for $i = 1, 2, \ldots, n$, set up the design matrix **D** and find the least squares estimates of α and β. Verify the formulas in some elementary statistics book or in Chapter 10.

6.7* (*Generalized inverses*) In the text of Chapter 5, it was mentioned that without constraints on the α_i and β_j, they were not uniquely determined and therefore could not be estimated. But need the parameters be estimated to find the projections onto the subspaces? Might not the appropriate sums of squares be calculated without recourse to specific bases or parameter estimates? This type of approach may be made using the concept of a *generalized inverse*. This is made explicit in the following subproblems.

(a) *DEFINITION* A *generalized inverse* of a matrix **M** is any matrix **G** for which $\mathbf{MGM} = \mathbf{M}$. Note that **M** is not assumed to be a square matrix. Show by example that **G** need not be unique.

(b) *Existence of the generalized inverse*:

(i) Prove the following theorem. Let **M**, an $m \times n$ matrix, have rank r. There exist nonsingular square matrices **A** and **B** such that

$$\mathbf{AMB} = \begin{pmatrix} \mathbf{I}^{r \times r} & \mathbf{0}^{r \times (n-r)} \\ \mathbf{0}^{(m-r) \times r} & \mathbf{0}^{(m-r) \times (n-r)} \end{pmatrix}.$$

[*Hint*: Perform the echelon reduction of **M** to diagonal form with 0's and 1's on the diagonal. Note that the operations are equivalent to pre- or post-multiplication by square matrices.]

(ii) Show that

$$\mathbf{G} = \mathbf{B}\begin{pmatrix} \mathbf{I}^{r \times r} & \mathbf{0}^{r \times (m-r)} \\ \mathbf{0}^{(n-r) \times r} & \mathbf{0}^{(n-r) \times (m-r)} \end{pmatrix}\mathbf{A}$$

is a generalized inverse.

(c) Consider a generalized inverse **G** of **D′D**, where **D** is the design matrix of this section. Show that

$$\mathbf{D'DGD'} = \mathbf{D'}.$$

[*Outline of the proof*: (i) $(\mathbf{D'DGD'})\mathbf{D} = \mathbf{D'}(\mathbf{D})$.

(ii) If $\mathbf{AD'D} = \mathbf{BD'D}$, then $\mathbf{AD'} = \mathbf{BD'}$. [*Hint*: **AD′** and **BD′** have rows that lie in the subspace generated by columns of **D**. They have the same inner product with each column of **D**.]

(iii) Let $\mathbf{A} = \mathbf{D'DGD'}$, $\mathbf{B} = \mathbf{I}$.]

(d) The projection of **X** onto the subspace generated by the columns of **D** is **DGD′X**. [*Hint*: You must show

(i) that $\mathbf{X} - \mathbf{DGD'X}$ is perpendicular to the columns of **D** or that $\mathbf{D'}(\mathbf{X} - \mathbf{DGD'X}) = \mathbf{0}$, and

(ii) that **DGD′X** lies in the subspace generated by the columns of **D**.]

6.8* (*More on generalized inverses*) Prove the following:

(a) Let the equation $\mathbf{Ax} = \mathbf{y}$ be consistent. Then for any generalized inverse **G** of **A**, $\mathbf{x} = \mathbf{Gy}$ is a solution.

(b) The most general solution for **x** is

$$\mathbf{x} = \mathbf{Gy} + (\mathbf{GA} - \mathbf{I})\mathbf{z}$$

for arbitrary **z**.

6.9 (*Random effects continued*) In Problem 4.6 we considered a random effects one-way analysis of variance. In a two-way analysis of variance, both factors may be random. A balanced two-way layout with interaction will be assumed:

$$X_{ijk} = \mu + \alpha_i + \beta_j + \gamma_{ij} + e_{ijk},$$
$$i = 1, 2, \ldots, I, \qquad j = 1, 2, \ldots, J, \qquad k = 1, \ldots, K.$$

TABLE 6.6

SS	df	$E(MS)$
SS_A	$I-1$	$\sigma^2 + K\sigma_\gamma^2 + JK\sigma_A^2$
SS_B	$J-1$	$\sigma^2 + K\sigma_\gamma^2 + IK\sigma_B^2$
SS_{int}	$(I-1)(J-1)$	$\sigma^2 + K\sigma_\gamma^2$
SS_e	$IJ(K-1)$	σ^2

The α_i's, β_j's, γ_{ij}'s, and e_{ijk}'s are all mutually independent normal random variables with mean zero and variances σ_A^2, σ_B^2, σ_γ^2, σ^2 respectively. (See Scheffé,[1] Chapter 7, for a derivation or justification of such a model.)

(a) Show that the ANOVA table is as shown in Table 6.6.

(b) What is an appropriate F-test for no A main effect; that is, $\sigma_A^2 = 0$? Is this the same test as that used in the fixed effects analysis of variance?

6.10 (*Random effects continued*) In many situations an analysis of variance may have both fixed effect(s) and random effect(s) present. An example of this might be the example of Problem 4.6 with sex used as a second factor. In this case, the selected schools would be levels of the randomly selected random effect and sex would give a fixed effect at two levels. Models with both fixed and random effects are called *mixed* models. In the case of the balanced two-layout, an appropriate model is given by (factor A fixed, B random)

$$X_{ijk} = \mu + \alpha_i + \beta_j + \gamma_{ij} + e_{ijk},$$
$$i = 1, \ldots, I, \qquad j = 1, \ldots, J, \qquad k = 1, \ldots, K,$$

where $\sum_i \alpha_i = 0$, $\sum_i \gamma_{ij} = 0$ for all j, and β_j, γ_{ij}, and e_{ijk} are all normal with mean zero, and the e_{ijk} are $N(0, \sigma^2)$ independently of the β_j and γ_{ij}. The joint distribution of the β_j and γ_{ij} is given in terms of an $I \times I$ covariance matrix $\mathbf{C}$ with elements $c_{ii'}$. Then

$$\operatorname{cov}(\beta_j, \beta_{j'}) = \delta_{j,j'}\bar{c}_{\cdot\cdot},$$
$$\operatorname{cov}(\gamma_{ij}, \gamma_{i'j'}) = \delta_{j,j'}(c_{i,i'} - \bar{c}_{i\cdot} - \bar{c}_{\cdot i'} + \bar{c}_{\cdot\cdot}),$$
$$\operatorname{cov}(\beta_j, \gamma_{ij}) = \delta_{j,j'}(\bar{c}_{i\cdot} - \bar{c}_{\cdot\cdot}),$$

where δ is the Kronecker δ. (See Scheffé,[1] Chapter 8, for a discussion of this model.)

(a) Derive the ANOVA table shown in Table 6.7, where

$$\sigma_\gamma^2 = \textstyle\sum_i \operatorname{var}(\gamma_{ij})/(I-1), \; \sigma_B^2 = var(\beta_j).$$

(b) How do you test for the A main effect? The B main effect? Interaction? Are these the same tests as in the fixed effects balanced two-way analysis of variance?

[1] H. Scheffé, *The Analysis of Variance*, Wiley, New York, 1959.

TABLE 6.7

SS	df	E(MS)
SS_A	$I-1$	$\sigma^2 + K\sigma_\gamma^2 + JK\sum_i \alpha_i^2/(I-1)$
SS_B	$J-1$	$\sigma^2 + IK\sigma_B^2$
SS_{int}	$(I-1)(J-1)$	$\sigma^2 + K\sigma_\gamma^2$
SS_e	$IJ(K-1)$	σ^2

6.11 Consider a study designed to test two drugs. The first drug is licensed in the United States but not in Europe. The second drug is licensed in Europe but not the United States. The first drug will be tested at n_1 hospitals and the second drug at n_2 hospitals. How would one take into account a hospital effect? There is no way to examine the interaction between hospitals and drugs, as each hospital will use only one drug. Such designs are called *nested* designs. Hospitals are said to be nested within drugs.

To simplify the analysis, assume there are I drugs and J hospitals used for each drug. Let there be K individuals tested within each hospital. Assume the response variable, say blood pressure, may be considered normally distributed. The nested model is

$$X_{ijk} = \mu + \alpha_i + \beta_{j(i)} + e_{ijk},$$
$$i = 1, \ldots, I, \qquad j(i) = 1, \ldots, J, \qquad k = 1, 2, \ldots, K,$$

where $\sum_{i=1}^{I} \alpha_i = 0$ and $\sum_{j=1}^{J} \beta_{j(i)} = 0$ for each i. A portion of the analysis of variance table is given in Table 6.8.

(a) Complete the table by discovering appropriate values for the sum of squares column and the expected mean square column.

(b) What is an obvious weakness of the design given here for comparing the drugs?

(c) Suppose all the drugs are being tested in the same country and that the hospitals are selected at random. Suppose the $\beta_{j(i)}$ are $N(0, \sigma_H^2)$ independently of each other and the e_{ijk}.

(i) Find the ANOVA table as in part (a). Let the hospital within drug sums of squares be summed to give SS_H.

TABLE 6.8

Source	df
Drug	I − 1
Hospitals within drug 1	$J-1$
⋮	⋮
Hospitals within drug I	$J-1$
Individuals within Hospitals within drug	$IJ(K-1)$

(ii) Show that the appropriate F-test for differing drug effects involves the drug SS and SS_H. Contrast this F-test with the appropriate F-test in part (a).

6.12 From the means of Table 6.3, estimate all parameters (except σ^2) of the three-way ANOVA; that is, estimate

$$\mu,\ \alpha_1,\ \alpha_2,\ \beta_1,\ \beta_2,\ \gamma_1,\ \gamma_2,$$
$$(\alpha\beta)_{11},\ (\alpha\beta)_{12},\ (\alpha\beta)_{21},\ (\alpha\beta)_{22},\ (\alpha\gamma)_{11},\ (\alpha\gamma)_{12},$$
$$(\alpha\gamma)_{21},\ (\alpha\gamma)_{22},\ (\beta\gamma)_{11},\ (\beta\gamma)_{12},\ (\beta\gamma)_{21},\ (\beta\gamma)_{22},$$
$$(\alpha\beta\gamma)_{111},\ (\alpha\beta\gamma)_{112},\ (\alpha\beta\gamma)_{121},\ (\alpha\beta\gamma)_{122},$$
$$(\alpha\beta\gamma)_{211},\ (\alpha\beta\gamma)_{212},\ (\alpha\beta\gamma)_{221},\ (\alpha\beta\gamma)_{222}.$$

You only need estimate the "correct" eight parameters and the rest follow easily. Why?

6.13 From the information given in Example 6.1 and Problem 6.12, construct the ANOVA table for Example 6.1.

6.14 [R. Altshuler and H. Kassinove, The effects of skill and chance instructional sets, schedule of reinforcement and sex on children's temporal persistence, *Child Development* 46, 258–262 (1975)][2] It has been suggested that the well-known partial reinforcement extinction effect may not apply to human subjects under conditions perceived as skill determined. Instead, there is evidence to suggest that superior resistance to extinction following partial reinforcement is restricted to situations perceived as independent of the subject's skill or ability. Altshuler and Kassinove investigated the effect of instructional set and schedule of reinforcement on temporal persistence in children. Fifth-grade pupils worked on a modified anagram task which required the formation of words from letter sets. Following a timed six-trial acquisition series, each child was given a "no time limit" persistence trial comprised of letters of low word-producing capability. The dependent variable was temporal persistence, measured by the total number of seconds the subject worked at the task before deciding to stop.

Forty-eight boys and 48 girls were randomly assigned to one of 16 groups. Depending on the particular experimental condition, the groups differed in (a) instruction set received (this is a test of skill in which a required number of words must be produced versus being told that the task you are about to do is one involving pure chance or luck in which one and only one special word must be produced), (b) schedule of reinforcement—subjects were given success or failure feedback after each of the six acquisition trials in accord with a preordained scheduled (0%, 33%, 66%, 100% of the six trials called successful and independent of actual performance), or (c) sex of subject. This resulted in a balanced $2 \times 4 \times 2$ factorial design.

It was hypothesized that (a) children given skill instructions would persist longer than children given chance instructions, and (b) under skill instructions persistence would be greatest for the continuously reinforced group, whereas under chance instructions persistence would be greatest for the partially reinforced groups (33 and 66%). Construct the ANOVA table from the data in Table 6.9 and determine which F values are significant at the .05 level.

TABLE 6.9

Means and Standard Deviations of Temporal Persistence (in sec) by Instructional Set, Schedule of Reinforcement, and Sex of Subject

Schedule of reinforcement		Skill set		Chance set	
		Male	Female	Male	Female
0%:	Mean	322.5	428.3	412.0	421.3
	SD	39.0	83.8	64.1	77.8
33%:	Mean	367.2	420.3	333.3	407.2
	SD	83.1	103.1	57.7	89.2
66%:	Mean	395.2	465.7	362.5	390.2
	SD	80.6	87.7	90.8	81.8
100%:	Mean	496.7	451.0	368.8	280.5
	SD	123.0	96.2	78.9	46.0

6.15 Consider the balanced two-way factorial design without interaction versus the balanced design with interaction. If there is only one replication at each possible combination of the levels of the factors, the assumption of zero interactions (often called the assumption of additivity, since $\mu_{ij} = \mu + \alpha_i + \beta_j$) cannot be tested by our usual F-test for interaction since the "interaction sum of squares" takes over the role of the error sum of squares, or, in other words, because there are no degrees of freedom left for an error sum of squares. However, if we impose some restrictions on the form the γ_{ij} interaction may take, certain tests for nonadditivity can be developed.

Tukey's test for nonadditivity assumes that $\gamma_{ij} = G\alpha_i\beta_j$ and basically tests that H_0: $G = 0$ versus H_1: $G \neq 0$. The following heuristic (but wrong!) approach leads to the correct answer and is an easy way to remember the test. Consider the model $\mu_{ij} = \mu + \alpha_i + \beta_j + G\alpha_i\beta_j$ with $\sum \alpha_i = 0$ and $\sum \beta_j = 0$. Ignore that this model is no longer linear in the parameters. Since $\sum_i \sum_j \mu G\alpha_i\beta_j = \sum_i \sum_j \alpha_i G\alpha_i\beta_j = \sum_i \sum_j \beta_j G\alpha_i\beta_j = 0$ the μ, α_i, β_j, and $G\alpha_i\beta_j$ subspaces are orthogonal, and independent estimates of the parameters can be made by projecting the data vector onto the orthogonal subspaces. Since we do not know the α_i and β_j's for the interaction subspace, let us replace them by their estimates $\hat{\alpha}_i$ and $\hat{\beta}_j$ (which are still in the interaction subspace). In other words, consider the model $\mu_{ij} = \mu + \alpha_i + \beta_j + G\hat{\alpha}_i\hat{\beta}_j$, which is linear in the parameters. Show that our estimate of G is as usual

$$\hat{G} = \frac{\sum_i \sum_j \hat{\alpha}_i \hat{\beta}_j X_{ij}}{\sum_i \hat{\alpha}_i^2 \sum_j \hat{\beta}_j^2}$$

and the interaction sum of squares is $SS_G = \sum_i \sum_j (\hat{G}\hat{\alpha}_i\hat{\beta}_j)^2$.

Even though SS_G is not a quadratic form in the observations, but the quotient of a sixth-degree polynomial by a fourth-degree polynomial, under the assumption of additivity SS_G/σ^2 has a central chi-square distribution with 1 df which is statistically independent of $(SS_e - SS_G)/\sigma^2$, which has a central chi-square distribution with $(I-1)(J-1)-1$ df. Calculate the test statistic

$$F = \frac{SS_G}{(SS_e - SS_G)/[(I-1)(J-1)-1]}$$

for the data for Problem 5.19. Test at the .05 level the hypothesis $G = 0$.

A similar heuristic method can be used for deriving a test for nonadditivity for the Latin Square (see Chapter 7) and certain other experimental designs that assume additivity but have only one replication at each possible combination of the levels of the factors.

7 THE LATIN SQUARE

Using the basic ideas we have developed so far, many clever experimental designs have been devised for specific purposes. In this chapter we give an example of one such design called a Latin Square design. Recall from the previous chapter that an experiment was complete if there was at least one observation for each possible combination of levels taken on by the factors involved. The Latin Square design will be our first example of an incomplete design or an incomplete layout. The basic idea behind the Latin Square design is that if certain assumptions may be made, then the number of observations necessary to analyze various effects may be reduced considerably. In particular, in this case, we assume a three-factor design with no two-factor or three-factor interactions. At a minimum, a complete design would of necessity have $I \times J \times K$ observations if the first factor took on I levels, the second factor J levels, and the third factor K levels. In this particular design, we shall assume that all factors take on the same number N of levels. A Latin Square design, instead of using N^3 observations, will allow the estimation to be done with N^2 observations. The estimates of the parameters associated with different factors may be estimated independently.

Suppose we have three factors A,B,C, all of which take on the same number N of levels. Further, suppose we know that the model is additive in that the mean, if the respective factors take on levels i, j, and k, is

$$\mu_{ijk} = \mu + \alpha_i + \beta_j + \gamma_k \qquad \text{with} \quad \sum_i \alpha_i = \sum_j \beta_j = \sum_k \gamma_k = 0.$$

(N.B.: The additivity assumption is crucial to the analysis and the results can be very misleading when interactions are actually present.)

DEFINITION 7.1 A Latin Square of size N is a set of N symbols arranged in an $N \times N$ matrix so that each symbol appears exactly once in each row and exactly once in each column (so each symbol appears N times).

(1) Latin Squares of all sizes exist, e.g.,

$$\begin{matrix} 1 & 2 & 3 & 4 & \cdots & N \\ N & 1 & 2 & 3 & \cdots & N-1 \\ N-1 & N & 1 & 2 & \cdots & N-2 \\ \vdots & \vdots & \vdots & \vdots & & \vdots \\ 2 & 3 & 4 & 5 & \cdots & 1 \end{matrix}$$

(2) From any square $N!(N-1)!$, new squares may be made from the $N!$ permutations of the rows and the $(N-1)!$ permutations of the columns (that leave the first column fixed).

TABLE 7.1[a]

		B			
		1	2	⋯	*N*
A	1	k_{11}	k_{12}	⋯	k_{1N}
	2	k_{21}	k_{22}	⋯	k_{2N}
	⋮	⋮	⋮		⋮
	N	k_{N1}	k_{N2}	⋯	k_{NN}

[a] Latin Square entries give the k level of factor C.

We observe A,B,C at levels i,j,k in N^2 combinations given in Table 7.1. We observe A at level 1, B at level 1, and C at level k_{11}, etc. Thus, each level of A is observed once with each level of B and once with each level of C. As before, $\bar{X}_{ij.}$ is the average over the *one* observation $X_{ijk_{ij}}$. Then

$$\begin{aligned} \sum_i \sum_j X^2_{ijk_{ij}} &= \sum_i \sum_j (X_{ijk_{ij}} - \bar{X}_{i..} - \bar{X}_{.j.} - \bar{X}_{..k_{ij}} + 2\bar{X}_{...} \\ &\quad + \bar{X}_{i..} - \bar{X}_{...} + \bar{X}_{.j.} - \bar{X}_{...} + \bar{X}_{..k_{ij}} - \bar{X}_{...} + \bar{X}_{...})^2 \\ &= \underbrace{N^2\bar{X}^2_{...}}_{\substack{SS_{GM} \\ \text{Rank} = 1}} + \underbrace{N\sum_i (\bar{X}_{i..} - \bar{X}_{...})^2}_{\substack{SS_A \\ \text{Rank} = N-1}} \\ &\quad + \underbrace{N\sum_j (\bar{X}_{.j.} - \bar{X}_{...})^2}_{\substack{SS_B \\ \text{Rank} = N-1}} + \underbrace{N\sum_k (\bar{X}_{..k} - \bar{X}_{...})^2}_{\substack{SS_C \\ \text{Rank} = N-1}} \\ &\quad + \underbrace{\sum_{i,j} (X_{ijk_{ij}} - \bar{X}_{i..} - \bar{X}_{.j.} - \bar{X}_{..k_{ij}} + 2\bar{X}_{...})^2}_{\substack{SS_e \\ \text{Rank} \le (N-1)(N-2)}}. \end{aligned}$$

Note: For Rank $\leq (N-1)(N-2)$:

$$\sum_j (X_{ijk_{ij}} - \bar{X}_{i..} - \bar{X}_{.j.} - \bar{X}_{..k_{ij}} + 2\bar{X}_{...})$$
$$= N(\bar{X}_{i..} - \bar{X}_{i..} - \bar{X}_{...} - \bar{X}_{...} + 2\bar{X}_{...})$$
$$= 0 \qquad \text{and similarly for the } i.$$

Using our usual approach, we set rank $SS_e \leq (N-1)^2$. Unfortunately, this is too large under the present circumstances. Perhaps the easiest approach is to *know* that Cochran's theorem holds and to find the degrees of freedom for SS_e by subtraction; that is,

$$\text{Rank } SS_e = N^2 - 1 - (N-1) - (N-1) - (N-1)$$
$$= N^2 - 3N + 2 = (N-1)(N-2).$$

To see that Cochran's theorem holds it is convenient to have a little more theory.

LEMMA Let $\mathbf{L}_1, \ldots, \mathbf{L}_k$ be $n \times n$ matrices, $\mathbf{x}^{n \times 1}$ and for all $\mathbf{x}$, $\mathbf{x} = \mathbf{L}_1\mathbf{x} + \cdots + \mathbf{L}_k\mathbf{x}$. Suppose $(\mathbf{L}_j\mathbf{y})'(\mathbf{L}_i\mathbf{x}) = 0$ for $i \neq j$ and all $\mathbf{x}$ and $\mathbf{y}$. Then

(1) $\sum_j \text{Rank } \mathbf{L}_j = n$.
(2) $\|\mathbf{x}\|^2 = \|\mathbf{L}_1\mathbf{x}\|^2 + \cdots + \|\mathbf{L}_k\mathbf{x}\|^2$.
(3) By the proof of Cochran's theorem, the $\|\mathbf{L}_j\mathbf{x}\|^2$ are squared norms of projections onto orthogonal subspaces (which generate all of R^n).

Proof As $\mathbf{y}'\mathbf{L}_j'\mathbf{L}_i\mathbf{x} = 0$ $(i \neq j)$, $\mathbf{L}_j'\mathbf{L}_i = \mathbf{0}^{n \times n}$. Thus the non-zero columns of the different $\mathbf{L}_j$'s are orthogonal and thus linearly independent. Choosing bases for the subspaces spanned by the columns of the $\mathbf{L}_j$, the union is linearly independent and thus $\sum_j \text{Rank } \mathbf{L}_j \leq n$. Also note $\text{Rank } \mathbf{L}_j'\mathbf{L}_j = \text{Rank } \mathbf{L}_j$.

(2) is immediate as $\mathbf{x}'\mathbf{x} = (\mathbf{x}'\mathbf{L}_1' + \cdots + \mathbf{x}'\mathbf{L}_k')(\mathbf{L}_1\mathbf{x} + \cdots + \mathbf{L}_k\mathbf{x}) = \mathbf{x}'\mathbf{L}_1'\mathbf{L}_1\mathbf{x} + \cdots + \mathbf{x}'\mathbf{L}_k'\mathbf{L}_k\mathbf{x} = \|\mathbf{L}_1\mathbf{x}\|^2 + \cdots + \|\mathbf{L}_k\mathbf{x}\|^2$.

From $\sum_j \text{Rank } \mathbf{L}_j \leq n$ and (2), (3) follows and thus (1). □

For the Latin Square,

$$X_{ijk_{ij}} = \underbrace{X_{ijk_{ij}} - \bar{X}_{i..} - \bar{X}_{.j.} - \bar{X}_{..k_{ij}} + 2\bar{X}_{...}}_{(\mathbf{L}_1\mathbf{X})_{ij}}$$
$$+ \underbrace{\bar{X}_{i..} - \bar{X}_{...}}_{(\mathbf{L}_2\mathbf{X})_{ij}} + \underbrace{\bar{X}_{.j.} - \bar{X}_{...}}_{(\mathbf{L}_3\mathbf{X})_{ij}} + \underbrace{\bar{X}_{..k_{ij}} - \bar{X}_{...}}_{(\mathbf{L}_4\mathbf{X})_{ij}} + \underbrace{\bar{X}_{...}}_{(\mathbf{L}_5\mathbf{X})_{ij}}.$$

It is easy to check that all cross-products $(\mathbf{L}_i\mathbf{y})'(\mathbf{L}_j\mathbf{x}) = 0$. Two examples follow:

$$(\mathbf{L}_2\mathbf{y})'(\mathbf{L}_3\mathbf{x}) = \sum_i \sum_j (\bar{y}_{i..} - \bar{y}_{...})(\bar{x}_{.j.} - \bar{x}_{...})$$
$$= \sum_i (\bar{y}_{i..} - \bar{y}_{...}) \sum_j (\bar{x}_{.j.} - \bar{x}_{...}) = 0 \cdot 0 = 0;$$

$$(\mathbf{L}_1\mathbf{y})'(\mathbf{L}_4\mathbf{x}) = \sum_i \sum_k (y_{ij_{i,k}k} - \bar{y}_{i..} - \bar{y}_{.j_{i,k}.} - \bar{y}_{..k} + 2\bar{y}_{...})(\bar{x}_{..k} - \bar{x}_{...})$$

think of j
as a function
of i and k!

$$= \sum_k ((\bar{x}_{..k} - \bar{x}_{...}) \sum_i (y_{ij_{i,k}k} - \bar{y}_{i..} - \bar{y}_{.j_{i,k}.} - \bar{y}_{..k} + 2\bar{y}_{...}))$$

$$= \sum_k (\bar{x}_{..k} - \bar{x}_{...})N(\bar{y}_{..k} - \bar{y}_{...} - \bar{y}_{...} - \bar{y}_{..k} + 2\bar{y}_{...})$$

$$= 0$$

Thus, Cochran's theorem applies and

$$N^2 = \sum \text{Rank SS's} = \text{Rank SS}_e + (N-1) + (N-1) + (N-1) + 1$$

or

$$\text{Rank SS}_e = N^2 - 3N + 2 = (N-1)(N-2).$$

Thus the Latin Square ANOVA table is given by Table 7.2.

TABLE 7.2

Latin Square ANOVA Table

Source	SS	df	E(MS)
A	$N\sum_i(\bar{X}_{i..} - \bar{X}_{...})^2$	$N-1$	$\sigma^2 + N\sum_i \alpha_i^2/(N-1)$
B	$N\sum_j(\bar{X}_{.j.} - \bar{X}_{...})^2$	$N-1$	$\sigma^2 + N\sum_j \beta_j^2/(N-1)$
C	$N\sum_k(\bar{X}_{..k} - \bar{X}_{...})^2$	$N-1$	$\sigma^2 + N\sum_k \gamma_k^2/(N-1)$
Error	$\sum_i\sum_j(X_{ijk_{ij}} - \bar{X}_{i..} - \bar{X}_{.j.} - \bar{X}_{..k_{ij}} + 2\bar{X}_{...})^2$	$(N-1)(N-2)$	σ^2
Total	$\sum_i\sum_j(X_{ijk_{ij}} - \bar{X}_{...})^2$	N^2-1	

The Latin Square design can be considered as an extension of the randomized block design. While the randomized block design is employed to minimize the effects of one nuisance variable in the comparison of treatment effects, the Latin Square design extends this principle to a double blocking that eliminates nuisance variables in two "directions" (a row direction and column direction) from experimental error. In the Latin Square design, the treatment levels are assigned randomly to the experimental units with the restriction that each treatment level appear in only one row block and one column block according to the Latin Square design. A restriction which clearly limits the use of a Latin Square design in its simple form is that the number of row blocks, the number of column blocks, and the number of treatment levels must all be equal.

Latin Square designs are often used in:

(1) Agricultural field experiments in which one is worried about fertility gradients in several directions.

(2) "Blocking" simultaneously on two factors. For example, the Latin Square design is used frequently in medical and psychological experiments in which the same experimental unit receives all the treatments in succession when there is no carry-over of the treatment effect from one period to another. Such a design is useful when there is a systematic effect of the order in which the treatments are given or when there is a common additive time trend between periods for the response variable. In this example, each experimental unit is a block and each time period is a block, so that the design tries to eliminate differences between experimental units and order effects (or trends) from the experimental error.

Example 7.1 [H. W. Brown, B. G. Harmon, and A. H. Jensen, Total sulfur-containing amino acids, isoleucine, and tryptophan requirements of the finishing pig for maximum nitrogen retention, *Journal of Animal Science* **38**, No. 1 (1974)] Brown *et al.* used three 4 × 4 Latin Square designs to study the total sulfur-bearing amino acids, isoleucine, and tryptophan requirements of the finishing pig for maximum nitrogen retention. In trial 1 supplemental DL-methionine, in trial 2 L-isoleucine, and in trial 3 L-

TABLE 7.3

Nitrogen Retention by Finishing Barrows Fed Graded Levels of DL-*Methionine* (*Trial 1*)

Total sulfur amino acids (% of diet)	Period	Barrow	Daily nitrogen retention (g)
.17	*A*	2	16.01
	B	1	11.99
	C	3	14.66
	D	4	18.21
.26	*A*	3	17.27
	B	2	18.03
	C	4	14.40
	D	1	12.76
.35	*A*	1	12.33
	B	4	15.99
	C	2	13.59
	D	3	16.08
.44	*A*	4	17.01
	B	3	15.40
	C	1	14.74
	D	2	16.31

tryptophan were substituted isonitrogenously for L-glutamic acid to provide, respectively, the levels of total sulfur-bearing amino acids, isoleucine, and tryptophan shown in Tables 7.3, 7.12, and 7.13. (The last two tables are in the problem section.) In each trial the 4 × 4 Latin Square used four finishing barrows (castrated pigs), four levels of the amino acid, and four 10-day periods (periods were consecutive and consisted of a 5-day pretest portion followed by a 5-day collection period).

Data showing the response to graded levels of methionine are shown in Table 7.3. Supplementing the basal diet, which assayed .17% total sulfur-bearing amino acids with DL-methionine, did not significantly increase nitrogen retention. Therefore, the total sulfur-bearing amino acids requirement estimated from this trial was .17% (the lowest level tested) or less. There were no significant differences in nitrogen retention among the four periods.

Example 7.2 [T. Eden and R. A. Fisher, Studies in crop variation. VI. Experiments on the response of the potato to potash and nitrogen, *Journal of Agricultural Science* **19**, Part II, 201–213 (1929)] Latin Square designs are often used in agricultural experiments because of fertility gradients due to geographic location. The following example is one of the earliest uses of Latin Squares in this context.

The investigators were interested in potash manuring. The potash was applied in three different forms: sulfate (S), muriate (M), and potash manure salts (P), and a control treatment of no potash (O). Of the 576 possible alternative 4 × 4 Latin Squares, one was selected at random for a 1925 study. The resulting data with yields in pounds per plot are given in Table 7.4; the row, column, and treatment sums are presented in Table 7.5; the analysis of variance table is shown in Table 7.6. At the .05 level, treatments is highly significant. The $\alpha = .05$ critical value is 4.76 for 3 and 6 df so that there was some type of differential effect in the column direction (thus demonstrating the need for a sophisticated design of this type.)

TABLE 7.4

M 444	P 422	O 173	S 398
O 279	S 439	M 423	P 409
P 436	M 428	S 445	O 212
S 453	O 237	P 410	M 393

TABLE 7.5

Rows		Columns		Treatments	
1	1437	1	1612	O	901
2	1550	2	1526	S	1735
3	1521	3	1451	M	1688
4	1493	4	1412	P	1677

TABLE 7.6

Source	SS	df	MS	F
Treatments	120,175	3	40,058	120.48
Rows	1740	3	580	1.74
Columns	5841	3	1947	5.86
Error	1995	6	332.5	—
Total	129,751	15		

The significant result for treatments is not surprising since one treatment was no potash. Given that the three treatments had an effect compared to no treatment, the crucial question is whether the three treatments S, M, and P differed among themselves. For this purpose, the treatment sum of squares was partitioned into two parts, one comparing potash versus no potash, and the other being the sum of squares among the three potash treatments. The partitioning of the sum of squares is given by

$$4\sum_k(\bar{X}_{\cdot\cdot k}-\bar{X}_{\cdot\cdot\cdot})^2 = 4\left\{\frac{3}{4}\left(\hat{\alpha}_O-\frac{\hat{\alpha}_S+\hat{\alpha}_P+\hat{\alpha}_M}{3}\right)^2+\left(\hat{\alpha}_S-\frac{\hat{\alpha}_S+\hat{\alpha}_P+\hat{\alpha}_M}{3}\right)^2\right.$$
$$\left.+\left(\hat{\alpha}_P-\frac{\hat{\alpha}_S+\hat{\alpha}_P+\hat{\alpha}_M}{3}\right)^2+\left(\hat{\alpha}_M-\frac{\hat{\alpha}_S+\hat{\alpha}_P+\hat{\alpha}_M}{3}\right)^2\right\},$$

where $\hat{\alpha}_S = \bar{X}_{\cdot\cdot S} - \bar{X}_{\cdot\cdot\cdot}$, etc.

The first term on the right-hand side has rank 1 and the last three terms add to a quadratic form of rank 2 (as it is essentially a sample variance of

TABLE 7.7

Analysis of Variance Table

Source	*SS*	*df*	*MS*	*F*
Treatments	120,175	3	40,058	120.48
Potash versus no potash	119,700	1	119,700	360
Potash treatments	475	2	237.5	0.711
Rows	1740	3	580	1.74
Columns	5841	3	1947	5.86
Error	1995	6	332.5	
Total	129,175	15		

three terms). One can thus apply Cochran's theorem, and the analysis of variance table may be written as shown in Table 7.7. From this we do not conclude that there is any difference among the three potash treatments since the $F_{2,6}$ value of .71 is not statistically significant.

Example 7.3 [W. J. Byford, Experiments with fungicide sprays to control *Ramularia beticola* in sugar-beet seed crops, *Annals of Applied Biology* **82**, 291–297 (1976)] Data kindly supplied by the author give the percentage of leaf spot on midstem leaves. The four treatments assigned to letters at random are: unsprayed, *B*; 1 spray, *C*; 2 sprays, *D*; 3 sprays, *A*. The column and row variables represent the physical layout of the plots. The data are given in Table 7.8; the ANOVA table is shown in Table 7.9. The treatment effects do have a significant *F* value. The next step is to see which means differ, a topic considered in Chapter 8.

TABLE 7.8

					Total
	B 11.071	*C* 2.426	*A* 1.611	*D* 1.216	16.324
	D 1.433	*A* 0.936	*B* 7.787	*C* 2.552	12.708
	A 1.045	*D* 2.442	*C* 3.956	*B* 9.136	16.579
	C 2.746	*B* 7.815	*D* 0.672	*A* 1.620	12.853
Total	16.295	13.619	14.026	14.524	58.464

Treatment totals: *A*, 5.212; *B*, 35.809; *C*, 11.680; *D*, 5.763

TABLE 7.9

ANOVA Table

Source	SS	df	MS	*F*	
Treatments	156.14	3	52.05	49.76	$p < .001$
Rows	3.38	3	1.13	1.08	
Columns	1.04	3	.35	.33	
Error	6.28	6	1.05		
Total	166.84	15			

PROBLEMS

7.1 The example from the paper by Eden and Fisher also had Latin Square data from the year 1926. The data are given in Table 7.10. Analyze

TABLE 7.10

M 584.0	S 557.0	O 461.5	P 498.5
S 519.5	P 485.5	M 477.0	O 389.0
P 474.5	O 378.5	S 467.5	M 491.5
O 464.0	M 511.0	P 507.0	S 492.0

the data as in the example and tell whether the same conclusions hold in the next year.

7.2 Another of Byford's *Ramularia* Latin Square designs (in 1971 at Pointon) gives data on seed yield (g/plot):

(a) Construct the ANOVA table for this Latin Square.

(b) Which F values are statistically significant at the .10, .05, or .01 levels?

TABLE 7.11

C 3009	*B* 2845	*D* 3492	*A* 3099
B 3166	*A* 3327	*C* 3588	*D* 3755
D 3721	*C* 3288	*A* 3878	*B* 3873
A 4056	*D* 3919	*B* 4120	*C* 3746

7.3 Show that the least squares estimates of the parameters are

$$\hat{\mu} = \bar{X}_{...}, \qquad \hat{\alpha}_i = \bar{X}_{i..} - \bar{X}_{...}, \qquad \hat{\beta}_j = \bar{X}_{.j.} - \bar{X}_{...}, \qquad \hat{\gamma}_k = \bar{X}_{..k} - \bar{X}_{...}.$$

7.4 Find the parameter estimates for:

(a) Example 7.1. (b) Example 7.2. (c) Example 7.3.

(d) Problem 7.1. (e) Problem 7.2.

7.5 (*Example 7.1 continued*) Data on nitrogen retention with L-isoleucine for Example 7.2 is given in Table 7.12. Construct an ANOVA table and analyze the data.

7.6 (*Example 7.1 continued*) The nitrogen retention data from the L-tryptophan trial is given in Table 7.13. Construct an ANOVA table and analyze the data.

TABLE 7.12

Nitrogen Retention by Finishing Barrows Fed Graded Levels of L-*Isoleucine* (*Trial* 2)

Total L-isoleucine (% of diet)	Period	Barrow	Daily nitrogen retention (g)
.22	*A*	3	15.97
	B	1	16.09
	C	4	10.75
	D	2	14.55
.30	*A*	4	12.33
	B	3	19.17
	C	2	15.77
	D	1	18.87
.38	*A*	2	16.21
	B	4	13.07
	C	1	19.21
	D	3	17.16
.46	*A*	1	18.49
	B	2	18.02
	C	3	17.71
	D	4	14.60

TABLE 7.13

Nitrogen Retention by Finishing Barrows Fed Graded Levels of L-*Tryptophan* (*Trial* 3)

Total L-tryptophan (% of diet)	Period	Barrow	Daily nitrogen retention (g)
.032	*A*	2	5.16
	B	3	13.56
	C	1	16.20
	D	4	12.24
.047	*A*	4	16.14
	B	1	17.14
	C	3	14.74
	D	2	13.01
.062	*A*	3	16.05
	B	4	17.22
	C	2	16.49
	D	1	21.33
.077	*A*	1	19.32
	B	2	18.07
	C	4	18.55
	D	3	18.77

8 CONFIDENCE SETS, SIMULTANEOUS CONFIDENCE INTERVALS, AND MULTIPLE COMPARISONS

In the previous chapters we have seen how to estimate the parameters associated with our analysis of variance models. However, the estimates by themselves are not very useful if we do not have an idea of how reliable the estimates are. For this reason we turn to the notion of confidence intervals and confidence sets. Our situation is typically complicated by having quite a few parameters that are simultaneously of interest. Some approaches to this problem are given in the remainder of this chapter for $\mathbf{X} \sim N(\mathbf{D}\boldsymbol{\beta}, \sigma^2\mathbf{I})$.

ONE-DIMENSIONAL CASE

For one unknown parameter, say $\mathbf{c}'\boldsymbol{\beta}$ ($\mathbf{c}$ a known $p \times 1$ vector), it is easy to find an appropriate $100(1-\alpha)\%$ confidence interval. Let $\mathrm{MS_e}$ denote the error mean square in our ANOVA table. Then $\hat{\boldsymbol{\beta}}$, the least squares estimate, and $\mathrm{MS_e}$ are independent as they involve projections onto orthogonal subspaces. As $\hat{\boldsymbol{\beta}} \sim N(\boldsymbol{\beta}, \sigma^2(\mathbf{D}'\mathbf{D})^{-1})$, $\mathbf{c}'\hat{\boldsymbol{\beta}} \sim N(\mathbf{c}'\boldsymbol{\beta}, \sigma^2\mathbf{c}'(\mathbf{D}'\mathbf{D})^{-1}\mathbf{c})$ is independent of $\mathrm{MS_e} = \mathrm{SS_e}/m$, where $\mathrm{SS_e}/\sigma^2 \sim \chi_m{}^2$. Thus,

$$\frac{\mathbf{c}'\hat{\boldsymbol{\beta}} - \mathbf{c}'\boldsymbol{\beta}}{\sigma\sqrt{\mathbf{c}'(\mathbf{D}'\mathbf{D})^{-1}\mathbf{c}}} \Bigg/ \sqrt{\frac{\mathrm{MS_e}}{\sigma^2}}$$

is a t random variable with m df (note that the σ cancels). An appropriate $100(1-\alpha)\%$ confidence interval for $\mathbf{c}'\boldsymbol{\beta}$ is given (as usual for a t random variable) by

$$(\mathbf{c}'\hat{\boldsymbol{\beta}} - t_{\alpha/2,m}\sqrt{\mathbf{c}'(\mathbf{D}'\mathbf{D})^{-1}\mathbf{c}}\sqrt{\mathrm{MS_e}},\ \mathbf{c}'\hat{\boldsymbol{\beta}} + t_{\alpha/2,m}\sqrt{\mathbf{c}'(\mathbf{D}'\mathbf{D})^{-1}\mathbf{c}}\sqrt{\mathrm{MS_e}}), \quad (1)$$

where $t_{\alpha/2,m}$ is the $\alpha/2$ critical value for the t-distribution with m df; that is, if T is a t random variable with m df, then

$$P(T \geq t_{\alpha/2,m}) = \alpha/2.$$

MULTIDIMENSIONAL CASE

Consider the case $p = 2$. Rather than an interval, we find a two-dimensional set that is constructed in such a way that the probability that the true $\boldsymbol{\beta}$ is in the set is $1 - \alpha$ (or $\geq 1 - \alpha$). In two dimensions the set will be an ellipsoid.

DEFINITION 8.1 A 100 $(1 - \alpha)\%$ confidence set for a parameter vector $\boldsymbol{\beta}$ is a random set, say $A(\mathbf{X})$ ($\mathbf{X}$ the data vector), such that $P(\boldsymbol{\beta}$ is in $A(\mathbf{X})) \geq 1 - \alpha$ for all possible $\boldsymbol{\beta}$ (when $\boldsymbol{\beta}$ is the true parameter vector). (Usually, the probability is equal to $1 - \alpha$ if we can arrange this.)

For our purposes we shall use the following lemma.

LEMMA 8.1 Let $\tilde{\boldsymbol{\beta}}^{p \times 1} \sim N(\boldsymbol{\beta},\mathbf{C})$, where $\mathbf{C}$ is positive definite. Then $(\tilde{\boldsymbol{\beta}} - \boldsymbol{\beta})'\mathbf{C}^{-1}(\tilde{\boldsymbol{\beta}} - \boldsymbol{\beta})$ has a central χ^2-distribution with p df.

Proof Using the Principal Axis theorem, there exists an orthogonal matrix $\mathbf{P}$ such that $\mathbf{PCP}' = \mathbf{D}$, where $\mathbf{D}$ is a diagonal matrix whose diagonal elements are the positive (why?) eigenvalues of $\mathbf{C}$. If

$$\mathbf{D} = \begin{pmatrix} d_1 & 0 & \cdots & 0 \\ 0 & d_2 & & 0 \\ \vdots & \vdots & \ddots & \vdots \\ 0 & 0 & \cdots & d_p \end{pmatrix},$$

let

$$\mathbf{D}^{-1/2} = \begin{pmatrix} 1/d_1^{1/2} & 0 & \cdots & 0 \\ 0 & 1/d_2^{1/2} & \cdots & 0 \\ \vdots & \vdots & & \vdots \\ 0 & 0 & \cdots & 1/d_p^{1/2} \end{pmatrix},$$

Then since $(\tilde{\boldsymbol{\beta}} - \boldsymbol{\beta}) \sim N(\mathbf{0}, \mathbf{C})$, we have

$$\begin{aligned} \mathbf{D}^{-1/2}\mathbf{P}(\tilde{\boldsymbol{\beta}} - \boldsymbol{\beta}) &\sim N(\mathbf{D}^{-1/2}\mathbf{P0}, \mathbf{D}^{-1/2}\mathbf{PC}(\mathbf{D}^{-1/2}\mathbf{P})') \\ &\sim N(\mathbf{0}, \mathbf{D}^{-1/2}\mathbf{PCP}'\mathbf{D}^{-1/2}) \sim N(\mathbf{0}, \mathbf{D}^{-1/2}\mathbf{DD}^{-1/2}) \sim N(\mathbf{0}, \mathbf{I}). \end{aligned}$$

Thus, $\|\mathbf{D}^{-1/2}\mathbf{P}(\tilde{\boldsymbol{\beta}} - \boldsymbol{\beta})\|^2 \sim \chi_p{}^2$. Now

$$\begin{aligned} \|\mathbf{D}^{-1/2}\mathbf{P}(\tilde{\boldsymbol{\beta}} - \boldsymbol{\beta})\|^2 &= (\tilde{\boldsymbol{\beta}} - \boldsymbol{\beta})'\mathbf{P}'\mathbf{D}^{-1/2}\mathbf{D}^{-1/2}\mathbf{P}(\tilde{\boldsymbol{\beta}} - \boldsymbol{\beta}) \\ &= (\tilde{\boldsymbol{\beta}} - \boldsymbol{\beta})'\mathbf{P}'\mathbf{D}^{-1}\mathbf{P}(\tilde{\boldsymbol{\beta}} - \boldsymbol{\beta}) = (\tilde{\boldsymbol{\beta}} - \boldsymbol{\beta})'\mathbf{C}^{-1}(\tilde{\boldsymbol{\beta}} - \boldsymbol{\beta}) \sim \chi_p{}^2. \end{aligned}$$

□

The proof looks mysterious until we think of the geometry that motivates it. The exponent of the normal density function contains the term $(\tilde{\boldsymbol{\beta}} - \boldsymbol{\beta})'\mathbf{C}^{-1}(\tilde{\boldsymbol{\beta}} - \boldsymbol{\beta})$. This being a quadratic form, we can rotate the space so that $\mathbf{C}$ (or $\mathbf{C}^{-1}$) is diagonal so that the density factors, and we have independent normal variables in the directions of this new coordinate system. The $\mathbf{D}^{-1/2}$ term divides each independent normal variable by its standard deviation in order to get $N(0, 1)$ random variables. The sum of their squares has a central χ^2-distribution. The only question remaining is to go back to the original coordinate system to see what our χ^2 variable looks like.

Returning to our least squares estimate $\hat{\boldsymbol{\beta}} \sim N(\boldsymbol{\beta}, \sigma^2(\mathbf{D}'\mathbf{D})^{-1})$, we have $(\hat{\boldsymbol{\beta}} - \boldsymbol{\beta})'(\mathbf{D}'\mathbf{D})(\hat{\boldsymbol{\beta}} - \boldsymbol{\beta})/\sigma^2$ is $\chi_p{}^2$. If the "nuisance" parameter σ^2 were not involved, we could use χ^2 critical values to get our confidence set; that is,

$$P(\boldsymbol{\beta} \text{ is in } \{\tilde{\boldsymbol{\beta}}:(\hat{\boldsymbol{\beta}} - \tilde{\boldsymbol{\beta}})'(\mathbf{D}'\mathbf{D})(\hat{\boldsymbol{\beta}} - \tilde{\boldsymbol{\beta}}) \leq \sigma^2\chi^2_{\alpha,p}\}) = 1 - \alpha$$

where $\chi^2_{\alpha,p}$ is such that if X is a χ^2 random variable with p df, $P(X \geq \chi^2_{\alpha,p}) = \alpha$. Thus, the confidence set is

$$\{\tilde{\boldsymbol{\beta}}:(\hat{\boldsymbol{\beta}} - \tilde{\boldsymbol{\beta}})'(\mathbf{D}'\mathbf{D})(\hat{\boldsymbol{\beta}} - \tilde{\boldsymbol{\beta}}) \leq \sigma^2\chi^2_{\alpha,p}\}.$$

We do not usually know σ^2, so we use the time-honored device of dividing by an appropriate estimate.

LEMMA 8.2 Let $\hat{\boldsymbol{\beta}}$ be the least squares estimate of $\boldsymbol{\beta}$, $\mathbf{D}$ of rank p, and MS_e the error mean square with m df. Then

$$(\hat{\boldsymbol{\beta}} - \boldsymbol{\beta})'\mathbf{D}'\mathbf{D}(\hat{\boldsymbol{\beta}} - \boldsymbol{\beta})/p\text{MS}_\text{e}$$

has a central F-distribution with p and m df.

Proof Use

$$(\hat{\boldsymbol{\beta}} - \boldsymbol{\beta})'\mathbf{D}'\mathbf{D}(\hat{\boldsymbol{\beta}} - \boldsymbol{\beta})/p\text{MS}_\text{e} = \frac{(\hat{\boldsymbol{\beta}} - \boldsymbol{\beta})'\mathbf{D}'\mathbf{D}(\hat{\boldsymbol{\beta}} - \boldsymbol{\beta})/\sigma^2 p}{\text{MS}_\text{e}/\sigma^2}. \quad \square$$

THEOREM (COROLLARY) 8.1 Let $\hat{\boldsymbol{\beta}}^{p \times 1}$ be the least squares estimate, $\mathbf{D}$ of rank p, and MS_e the error mean square with m df. Then a $100(1 - \alpha)\%$ confidence set for $\boldsymbol{\beta}$ is

$$\{\tilde{\boldsymbol{\beta}}:(\hat{\boldsymbol{\beta}} - \tilde{\boldsymbol{\beta}})'\mathbf{D}'\mathbf{D}(\hat{\boldsymbol{\beta}} - \tilde{\boldsymbol{\beta}})/p\text{MS}_\text{e} \leq F_{\alpha,p,m}\},$$

where $F_{\alpha,p,m}$ is the α critical value of the F-distribution with p and m df; that is, F is an $F_{p,m}$ random variable and $P(F \geq F_{\alpha,p,m}) = \alpha$.

These confidence sets are ellipsoids in p-dimensional space.

SIMULTANEOUS CONFIDENCE INTERVALS USING BONFERRONI'S INEQUALITY

For events $A_1, \ldots, A_k$. $P(A_1 \cup A_2 \cup \cdots \cup A_k) \leq \sum_{i=1}^k P(A_i)$. This inequality is sometimes called Bonferroni's inequality. Applied to a confidence interval situation, we have:

LEMMA 8.3 Let $I_i(\mathbf{X})$ be a $100(1 - \alpha_i)\%$ confidence interval for the parameter γ_i, $i = 1, \ldots, k$. The probability that all k parameters are simultaneously in their confidence intervals is greater than or equal to $1 - \sum_{i=1}^k \alpha_i$; that is,

$$P\left(\bigcap_{i=1}^{k} (\gamma_i \in I_i(\mathbf{X}))\right) \geq 1 - \sum_{i=1}^{k} \alpha_i.$$

Proof Let superscript c denote the complement of a set.

$$\begin{aligned} P\left(\bigcap_{i=1}^{k} (\gamma_i \in I_i(\mathbf{X}))\right) &= 1 - P\left[\left\{\bigcap_{i=1}^{k} (\gamma_i \in I_i(\mathbf{X}))\right\}^{c}\right] \\ &= 1 - P\left(\bigcup_{i=1}^{k} [(\gamma_i \in I_i(\mathbf{X}))^{c}]\right) \\ &\geq 1 - \sum_{i=1}^{k} P((\gamma_i \in I_i(\mathbf{X}))^{c}) \\ &\geq 1 - \sum_{i=1}^{k} \alpha_i. \quad \square \end{aligned}$$

The reason that simultaneous confidence intervals are desired is that it is difficult to visualize $\boldsymbol{\beta}$ in an ellipsoid but easy to think of β_1 in $[a_1, b_2]$, β_2 in $[a_2, b_2]$, etc.

Example 8.1 [G. Gey, R. Levy, L. Fisher, G. Pettet, and R. Bruce, Plasma concentration of procainamide and prevalence of exertional arrythmias, *Annals of Internal Medicine* **80**, 718–722 (1974)] Gey *et al.* studied cardiac patients who had considerable arrythmia (heart beat irregularities) on a maximal exercise treadmill test. A drug, oral procainamide, was studied on 23 patients. A variety of variables measured before and during a maximal exercise test were studied. Two exercise tests were run, one as a control and one an hour after taking the drug. Among the variables to be compared were heart rate (HR), systolic blood pressure (SP), diastolic blood pressure (DP), frequency of arrhythmias, severity of arrhythmia (severity index), rate of maximum oxygen uptake ($V_{O_2\,\max}$), the amount of functional aerobic impairment (FAI), and computer ST_B, a measure of ischemic (lack of oxygen) response on the electrocardiogram.

TABLE 8.1

Variables at Rest and Exercise before and after Oral Procainamide[a]

		Rest						Exercise							
	Procainamide plasma level, 1 hr	HR		SP (mm Hg)		DP (mm Hg)		HR maximum		SP maximum (mm Hg)		DP maximum (mm Hg)		Arrhythmia frequency	
		Control	1 hr	Control	1 hr	Control	1 hr	Control	1 hr	Control	1 hr	Control	1 hr	Control	1 hr
	(μg/100 ml)														
Number of patients:	23	23		23		23		23		23		23		23	
Mean:	5.99	73	87	129	118	81	81	171	170	187	168	85	76	105	38
±SD:	±1.33	±11	±13	±17	±11.8	±11	±9.2	±13.5	±14	±20.6	±20	±12	±10	±108	±69
t:		5.053		4.183		.3796		.9599		5.225		5.005		3.422	
P^b:		<.0015		<.0060		NS		NS		<.0015		<.015		<.0360	

									Computer ST_B					
	Severity index		$\dot{V}_{O_2\,max}$(ml/min)		FAI (%)		Rest		Maximum		Slope		Zero recovery	
	Control	1 hr	Control	1 hr	Control	1 hr	Control	1 hr	Control	1 hr	Control	1 hr	Control	1 hr
Number of patients:	23		22		23		22		22		22		23	
Mean:	12.9	4.9	33.2	33.0	12.9	13.5	0.036	.044	−.190	−.122	−2.31	−2.05	−.065	−.0302
±SD:	±3.0	±4.67	±5.8	±6.0	±12.5	±11.5	±0.044	±.051	±.126	±.095	±1.401	±1.29	±.0003	±.077
t:	5.870		.3852		.5253		.8861		3.915		1.132		4.320	
P^b:	<.0015		NS		NS		NS		<.0120		NS		<.0045	

[a] Dose, 15 mg per kilogram body weight; HR = heart rate; SP = systolic pressure; DP = diastolic pressure; $\dot{V}_{O_2\,max}$ = maximal oxygen consumption; FAI = functional aerobic impairment; ST_B = 100-beat averaged ST depression, from monitored CBs lead, taken 50 to 69 msec after nadir of S wave; Slope = $\Delta HR/\Delta ST_B$; t = paired t-test; NS = not significant.

[b] Probability multiplied by 15 to correct for multiple comparisons (Bonferroni's inequality correction).

Fifteen comparisons were to be made from the before and after (control and with the drug) situations. Further, there is no reason to assume that the tests are statistically independent. To deal with the multiple comparison problem, the p values of each test were multiplied by the number of comparisons being made (Bonferroni's inequality). Thus, significance at the overall .05 level is at the .05/15 level for an individual comparison. Fourteen of the 15 comparisons (tests) were presented in the table from the paper (Table 8.1). Note that even with the multiple testing there were statistically significant differences for many of the tests.

THE S METHOD OF SIMULTANEOUS CONFIDENCE INTERVALS

The S stands for Professor Henry Scheffé who first presented the method. The geometry of the idea is clear as outlined below.

Suppose we are 90% confident that $\boldsymbol{\beta}$ lies in an ellipsoid ξ ($p = 2$ for simplicity of the pictures). Then we are 90% confident, in fact more than 90% confident, that $\beta_1 \in I_1$ *and* $\beta_2 \in I_2$, where I_1 and I_2 are the intervals indicated in Figure 8.1.

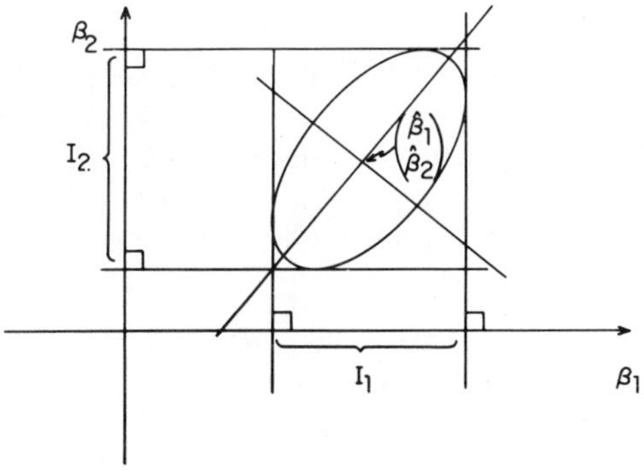

FIGURE 8.1

Thus, I_1 and I_2 give two simultaneous confidence intervals. Any closed *convex set* is determined by its *planes of support*. Looking at the figure, we see ways of getting other confidence statements (all simultaneously). For example, let $\mathbf{c}$ be a unit vector, that is, $\|\mathbf{c}\|^2 = 1$. Then $c_1\beta_1 + c_2\beta_2$ lies in the indicated interval I if $(\beta_1, \beta_2)'$ is in ξ (Figure 8.2). Note that $\mathbf{c}'\boldsymbol{\beta} = c_1\beta_1 + c_2\beta_2$ is the signed length of the projection of $\boldsymbol{\beta}$ onto the $\mathbf{c}$ direction (see Figure 8.3). If $\|\mathbf{c}\| \neq 1$, we can find a confidence interval for $\mathbf{c}'\boldsymbol{\beta}/\|\mathbf{c}\|$ and

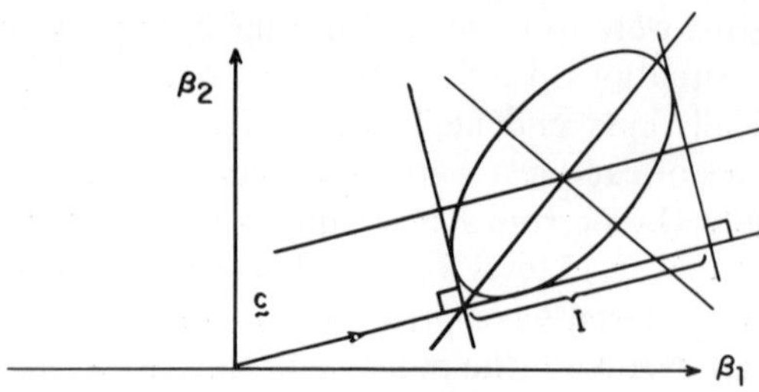

FIGURE 8.2

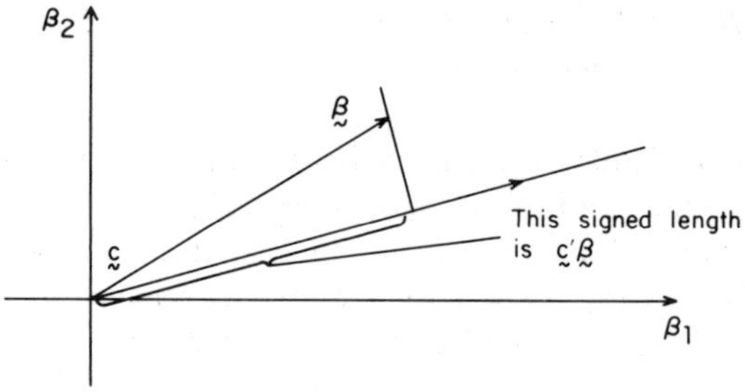

FIGURE 8.3

thus a confidence interval for $\mathbf{c}'\boldsymbol{\beta}$. Summarizing, since we have confidence ellipsoids for $\boldsymbol{\beta}$, it is geometrically clear that we can *simultaneously* get confidence intervals for all linear combinations of $\boldsymbol{\beta}$ with the given confidence. All that remains is to calculate analytically what the confidence intervals look like. The next theorem gives the result (stated so that it holds in a little more generality than just $\hat{\boldsymbol{\beta}}$).

THEOREM 8.2 Let $\tilde{\gamma}^{k\times 1} \sim N(\gamma, \sigma^2\mathbf{C})$, where $\sigma^2 > 0$ and $\mathbf{C}$ is positive definite. Independently of $\tilde{\gamma}$, let $ms^3/\sigma^2 \sim \chi_m{}^2$. Then with confidence, $1-\alpha$ *for all* $\mathbf{c}$,

$$\mathbf{c}'\gamma \text{ is in } \quad (\mathbf{c}'\tilde{\gamma} - \sqrt{\mathbf{c}'\mathbf{C}\mathbf{c}}\sqrt{kF_{\alpha,k,m}}s, \mathbf{c}'\tilde{\gamma} + \sqrt{\mathbf{c}'\mathbf{C}\mathbf{c}}\sqrt{kF_{\alpha,k,m}}s).$$

$F_{\alpha,k,m}$ is the α critical value of the F-distribution with k and m df.

Proof As in Lemma 8.2,

$$\frac{\|\mathbf{D}^{-1/2}\mathbf{P}(\tilde{\gamma} - \gamma)\|^2}{ks^2} \sim F_{k,m},$$

where $\mathbf{PCP}' = \mathbf{D}$ is diagonal and $\mathbf{P}$ orthogonal. Therefore

$$P(\|\mathbf{D}^{-1/2}\mathbf{P}(\tilde{\gamma} - \gamma)\| \le \sqrt{kF_{\alpha,k,m}}s) = 1 - \alpha.$$

The Cauchy–Schwarz inequality is $|\mathbf{a}'\mathbf{b}| \le \|\mathbf{a}\|\,\|\mathbf{b}\|$. Now

$$|\mathbf{c}'\tilde{\gamma} - \mathbf{c}'\gamma| = |\mathbf{c}'(\tilde{\gamma} - \gamma)| = |\underbrace{\mathbf{c}'\mathbf{P}'\mathbf{D}^{1/2}}_{\mathbf{a}}\underbrace{\mathbf{D}^{-1/2}\mathbf{P}(\tilde{\gamma} - \gamma)}_{\mathbf{b}}|$$

$$\le \|\mathbf{D}^{1/2}\mathbf{P}\mathbf{c}\|\,\|\mathbf{D}^{-1/2}\mathbf{P}(\tilde{\gamma} - \gamma)\|$$

Note that (Problem 8.2) $\|\mathbf{D}^{1/2}\mathbf{P}\mathbf{c}\| = \sqrt{\mathbf{c}'\mathbf{C}\mathbf{c}}$. □

COROLLARY 8.2 If $\mathbf{X} \sim N(\mathbf{D}\boldsymbol{\beta}, \sigma^2\mathbf{I})$ and $\mathbf{D}$ has rank p and $\hat{\boldsymbol{\beta}}$ is the least squares estimate of $\boldsymbol{\beta}$, then with confidence $1 - \alpha$ for all $\mathbf{d}^{p \times 1}$,

$$\mathbf{d}'\boldsymbol{\beta} \text{ is in}$$
$$(\mathbf{d}'\hat{\boldsymbol{\beta}} - \sqrt{\mathbf{d}'(\mathbf{D}'\mathbf{D})^{-1}\mathbf{d}}\sqrt{pF_{\alpha,p,m}\mathrm{MS_e}},\ \mathbf{d}'\hat{\boldsymbol{\beta}} + \sqrt{\mathbf{d}'(\mathbf{D}'\mathbf{D})^{-1}\mathbf{d}}\sqrt{pF_{\alpha,p,m}\mathrm{MS_e}})$$

when $\mathrm{MS_e}$ has m df.

Proof $\mathbf{d}'\hat{\boldsymbol{\beta}} = \mathbf{d}'(\mathbf{D}'\mathbf{D})^{-1}\mathbf{D}'\mathbf{X}$, so apply the theorem with $\mathbf{C} = \mathbf{I}$ and $\mathbf{c}' = \mathbf{d}'(\mathbf{D}'\mathbf{D})^{-1}\mathbf{D}'$. Note that $\mathbf{c}'E(\mathbf{X}) = \mathbf{d}'(\mathbf{D}'\mathbf{D})^{-1}\mathbf{D}'\mathbf{D}\boldsymbol{\beta} = \mathbf{d}'\boldsymbol{\beta}$. □

COROLLARY 8.3 For the preceding assumptions for all $\mathbf{c}$ in the estimation space, we have $100(1 - \alpha)\%$ confidence that

$$\mathbf{c}'\mathbf{D}\boldsymbol{\beta} \text{ is in } \quad (\mathbf{c}'\mathbf{X} - \|\mathbf{c}\|\sqrt{pF_{\alpha,p,m}\mathrm{MS_e}},\ \mathbf{c}'\mathbf{X} + \|\mathbf{c}\|\sqrt{pF_{\alpha,p,m}\mathrm{MS_e}}).$$

Proof Problem 8.3 □

COROLLARY 8.4 For the preceding assumptions, let S be a k-dimensional subspace of the estimation space of R^n. $\mathbf{X}^{n\times 1}N(\mathbf{D}\boldsymbol{\beta}, \sigma^2\mathbf{I})$. Then with confidence at least $1 - \alpha$ for all $\mathbf{c}$ in S,

$$\mathbf{c}'\mathbf{D}\boldsymbol{\beta} \text{ is in } \quad (\mathbf{c}'\mathbf{X} - \|\mathbf{c}\|\sqrt{kF_{\alpha,k,m}\mathrm{MS_e}},\ \mathbf{c}'\mathbf{X} + \|\mathbf{c}\|\sqrt{kF_{\alpha,k,m}\mathrm{MS_e}}).$$

Proof Let $\mathbf{c}_1, \ldots, \mathbf{c}_k$ be an orthonormal basis for S. Then

$$\tilde{\gamma} \equiv \begin{pmatrix} \mathbf{c}_1'\mathbf{X} \\ \vdots \\ \mathbf{c}_k'\mathbf{X} \end{pmatrix} \sim N(\begin{pmatrix} \mathbf{c}_1'\mathbf{D}\boldsymbol{\beta} \\ \vdots \\ \mathbf{c}_k'\mathbf{D}\boldsymbol{\beta} \end{pmatrix}, \sigma^2\mathbf{I}),$$

and Theorem 8.2 applies to $\tilde{\gamma}$. Let $\mathbf{c} \in S$. Then

$$\mathbf{c} = c_1\mathbf{c}_1 + \cdots + c_k\mathbf{c}_k \qquad \text{and} \qquad \mathbf{c}^{*\prime}\tilde{\gamma} \equiv (c_1, \ldots, c_k)\tilde{\gamma} = \mathbf{c}'\mathbf{X}$$

($\mathbf{c}^* \in R^k$ and $\mathbf{c} \in R^n$). Thus with probability $1 - \alpha$,

$$\mathbf{c}^{*\prime}\gamma = \mathbf{c}'\mathbf{D}\boldsymbol{\beta} \text{ is in } (\mathbf{c}^{*\prime}\tilde{\gamma} - \sqrt{\mathbf{c}^{*\prime}\mathbf{I}^{k\times k}\mathbf{c}^*}\sqrt{kF_{\alpha,k,m}\mathrm{MS_e}},\ \mathbf{c}^{*\prime}\tilde{\gamma} + \sqrt{\mathbf{c}^{*\prime}\mathbf{I}\mathbf{c}^*}\sqrt{kF_{\alpha,k,m}\mathrm{MS_e}}).$$

Now $\mathbf{c}^{*\prime}\tilde{\gamma} = \mathbf{c}'\mathbf{X}$ and $\mathbf{c}^{*\prime}\mathbf{I}\mathbf{c}^* = \sum_{i=1}^{k} c_i^2 = \|\mathbf{c}\|^2$, giving the desired result. □

T-*METHOD SIMULTANEOUS CONFIDENCE INTERVALS*

Professor John Tukey came up with another method of finding simultaneous confidence intervals. These are called T confidence intervals.

The T approach is based on the studentized range. Let $X_1, \ldots, X_k$ be i.i.d. $N(\mu, \sigma^2)$ random variables. Let s^2 be an independent estimate of σ^2 such that ms^2/σ^2 has a χ^2-distribution with m df. The random variable

$$Q_{k,m} = \frac{\max(X_1, \ldots, X_k) - \min(X_1, \ldots, X_k)}{s}$$

is called the studentized range. Denote the upper α critical value by $q_{\alpha,k,m}$. The distribution of $Q_{k,m}$ does not depend on the value of μ or σ^2 (Problem 8.5). The T method is given in the following theorem.

THEOREM 8.3 (a) Let $Z_1, \ldots, Z_k$ be $N(\mu_1, a\sigma^2), \ldots, N(\mu_k, a\sigma^2)$ respectively, and the Z_i's are multinormally distributed. Let $\text{cov}(Z_i, Z_j) = b\sigma^2$, $i \neq j$, where $b \leq 0$.

(b) Let s^2 be independent of $Z_1, \ldots, Z_k$, and $ms^2/\sigma^2 \sim \chi^2$ with m df.

Then the probability is $1 - \alpha$ that for all $\mathbf{c}' = (c_1, \ldots, c_k)$ with $\mathbf{c}'\mathbf{1} = 0$,

$$\mathbf{c}'\boldsymbol{\mu} \text{ is in } \left(\mathbf{c}'\mathbf{Z} - \sqrt{a-b}\,q_{\alpha,k,m}s\tfrac{1}{2}\sum_{i=1}^{k}|c_i|, \mathbf{c}'\mathbf{Z} + \sqrt{a-b}\,q_{\alpha,k,m}s\tfrac{1}{2}\sum_{i=1}^{k}|c_i|\right).$$

where $\mathbf{Z}' = (Z_1, \ldots, Z_k)$ and $\boldsymbol{\mu}' = (\mu_1, \ldots, \mu_k)$,

Proof (i) *It is enough to prove the theorem for uncorrelated* $Z_1, \ldots, Z_k$. Let W be $N(0, -b\sigma^2)$ and $Z_i' = Z_i + W$, W independent of $Z_1, \ldots, Z_k$. Then

$$\begin{aligned} \text{cov}(Z_i', Z_j') &= \text{cov}(Z_i, Z_j) + \text{cov}(W, W) = b\sigma^2 - b\sigma^2 = 0, \\ \text{var}(Z_i') &= \text{var}(Z_i) + \text{var}(W) = a\sigma^2 - b\sigma^2 = (a-b)\sigma^2. \end{aligned}$$

It is easily checked that the theorem for the Z_i' implies the theorem for the Z_i's as

$$\mathbf{c}'\begin{pmatrix} Z_1' \\ \vdots \\ Z_k' \end{pmatrix} = \mathbf{c}'\mathbf{Z} + \sum c_i W = \mathbf{c}'\mathbf{Z}.$$

(ii) *It is enough to prove the theorem for all* $\mathbf{c}'\boldsymbol{\mu} = \mu_i - \mu_{i'}$, $i \neq i'$. This will follow if we show that

$$\left|\sum_{i=1}^{k} c_i Z_i - \sum_{i=1}^{k} c_i \mu_i\right| \leq \tfrac{1}{2}\sum_{i=1}^{k}|c_i| \max_{i,i'}\left|(Z_i - \mu_i) - (Z_{i'} - \mu_{i'})\right|,$$

and then show that with probability $1 - \alpha$,

$$\max_{i,i'} |(Z_i - \mu_i) - (Z_{i'} - \mu_{i'})| \le s\sqrt{a - b}\, q_{\alpha,k,m}.$$

To see the first part, note that $\sum_{c_i > 0} c_i = -\sum_{c_i < 0} c_i = \frac{1}{2}\sum |c_i|$. Thus, if $\sum |c_i| > 0$, then

$$\frac{|\sum_{i=1}^k c_i Z_i - \sum_{i=1}^k c_i \mu_i|}{\frac{1}{2}\sum |c_i|} = \left| \frac{\sum_{ci>0} |c_i|(Z_i - \mu_i)}{\frac{1}{2}\sum |c_i|} - \frac{\sum_{ci<0} |c_i|(Z_i - \mu_i)}{\frac{1}{2}\sum |c_i|} \right|$$

But if we average a function over two different sets and subtract the averages, the difference is $\le$ the range of the function $= \max(Z_i - \mu_i) - \min(Z_i - \mu_i) = \max_{i,i'}|(Z_i - \mu_i) - (Z_{i'} - \mu_{i'})|$ which gives the inequality.

(iii)
$$\frac{\max_{i,i'}|(Z_i - \mu_i) - (Z_{i'} - \mu_{i'})|}{\sqrt{a-b}s} = \max_i \frac{(Z_i + W - \mu_i)}{\sqrt{a-b}s} - \min_i \frac{(Z_i + W - \mu_i)}{\sqrt{a-b}s}$$
$$= \frac{\max Y_i}{s} - \frac{\min Y_i}{s},$$

where $Y_i = (Z_i + W - \mu_i)/\sqrt{a - b}$ are $N(0, \sigma^2)$ variables which are independent as they are uncorrelated and normal. But by definition,

$$\frac{\max Y_i - \min Y_i}{s} = Q_{k,m} \le q_{\alpha,k,m}$$

with probability $1 - \alpha$. □

Note the following restrictions upon the T method:

(1) All the variances of the Z_i's must be the same.
(2) The covariances must have the same negative value.

(1) and (2) usually mean that the method may be used only in the case of balanced designs (as in unbalanced designs the variances of different estimates usually differ).

DEFINITION 8.2 A *contrast* of parameters

$$\boldsymbol{\beta} = \begin{pmatrix} \beta_1 \\ \beta_2 \\ \vdots \\ \beta_p \end{pmatrix}$$

is a sum $\sum_{i=1}^p c_i \beta_i$, where $\sum_{i=1}^p c_i = 0$. A *simple contrast* has one $c_i = 1$, one $c_{i'} = -1$, and the remainder equal to zero.

When conditions are such that both Tukey and Scheffé methods may be used, the following rule of thumb is useful. If only simple contrasts are being studied, use the T method; otherwise, use the S method. This will (approximately) give the shorter confidence intervals.

In order to use the Scheffé and Tukey multiple comparison methods, we need critical values for the F-distribution and the studentized range. Let us work out some examples of these multiple comparison procedures.

Example *(One-way analysis of variance)*

$$X_{ij} \sim N(\mu_i, \sigma^2), \qquad i = 1, \ldots, I, \quad j = 1, \ldots, J, \quad \text{all independent.}$$

(a) Find S-simultaneous $(1 - \alpha)100\%$ confidence intervals for all the μ_i. It is easy to show that

$$\hat{\mu}_i = \bar{X}_{i\cdot} \qquad \text{and} \qquad \sigma^2(\mathbf{D}'\mathbf{D})^{-1} = \sigma^2 \begin{pmatrix} 1/J & 0 & \cdots & 0 \\ 0 & & & \vdots \\ \vdots & & \ddots & \\ 0 & 0 & \cdots & 1/J \end{pmatrix}$$

Using $\mathbf{c}'\mathbf{D}\boldsymbol{\beta} = \mu_i$, where

$$\mathbf{c}' = (0, \ldots, 0, \underbrace{1/J, \ldots, 1/J}_{J \text{ position}}, 0, \ldots, 0),$$

and Corollary 8.3, with probability $1 - \alpha$, we see that all μ_i are in

$$\left(\bar{X}_{i\cdot} - \frac{1}{\sqrt{J}}\sqrt{IF_{\alpha,I,I(J-1)}\mathrm{MS_e}},\ \bar{X}_{i\cdot} + \frac{1}{\sqrt{J}}\sqrt{IF_{\alpha,J,I(J-1)}\mathrm{MS_e}}\right),$$

where $\mathrm{MS_e} = \sum_{i=1}^{I}\sum_{j=1}^{J}(X_{ij} - \bar{X}_{i\cdot})^2/I(J-1)$.

(b) Find the T-simultaneous confidence intervals for all the simple contrasts $\mu_i - \mu_j$, $i \neq j$.

Let $Z_i = \bar{X}_{i\cdot}$, $\operatorname{var}(\bar{X}_{i\cdot}) = (1/J)\sigma^2$, $\operatorname{cov}(\bar{X}_{i\cdot}, \bar{X}_{j\cdot}) = 0$, $i \neq j$. The $(1 - \alpha)100\%$ simultaneous confidence intervals are

$$\left(\bar{X}_{i\cdot} - \bar{X}_{j\cdot} - \frac{1}{\sqrt{J}} q_{\alpha,I,I(J-1)}\sqrt{\mathrm{MS_e}},\ \bar{X}_{i\cdot} - \bar{X}_{j\cdot} + \frac{1}{\sqrt{J}} q_{\alpha,I,I(J-1)}\sqrt{\mathrm{MS_e}}\right)$$

(here $\frac{1}{2}\sum|c_i| = \frac{1}{2}(1 + 1) = 1$).

(c) Find the S-confidence intervals for all the simple contrasts $\mu_i - \mu_j$, $i \neq j$. We first note that *all contrasts* $\sum c_i\mu_i$, $\sum c_i = 0$ *lie in the* $(I - 1)$-*dimensional space of* R^I *spanned by* $\mu_i - \mu_{i+1}$, $i = 1, 2, \ldots, I - 1$.

Proof of statement (a) We can generate $\mu_i - \mu_j$ (for any i and j). Without loss of generality, let $j = i + k$. Then

$$\mu_i - \mu_j = (\mu_i - \mu_{i+1}) + (\mu_{i+1} - \mu_{i+2}) + \ldots + (\mu_{i+k-1} - \mu_{i+k}).$$

(b) "Identifying" $\mu_i - \mu_j$ with

$$\begin{pmatrix} 0 \\ \vdots \\ 0 \\ 1 \\ 0 \\ \vdots \\ 0 \\ -1 \\ 0 \\ \vdots \\ 0 \end{pmatrix} \begin{matrix} \\ \\ \\ \leftarrow i\text{th position} \\ \\ \\ \\ \leftarrow j\text{th position} \\ \\ \\ \\ \end{matrix}$$

we see that $\mu_1 - \mu_2, \ldots, \mu_{I-1} - \mu_I$ are linearly independent and perpendicular to

$$\mathbf{1}^{I \times 1} = \begin{pmatrix} 1 \\ \vdots \\ 1 \end{pmatrix},$$

and thus span the orthogonal complement of the vector space generated by **1**. □

We may generate our S-type intervals by using the $I - 1$ functions $\bar{X}_{i\cdot} - \bar{X}_{i+1\cdot}$. Thus, by Corollary 8.4, for any contrast $\mathbf{c}'\boldsymbol{\mu}$ the simultaneous $1 - \alpha$ intervals are

$$\left(\mathbf{c}'\begin{pmatrix} \bar{X}_{1\cdot} \\ \vdots \\ \bar{X}_{I\cdot} \end{pmatrix} - \frac{\sqrt{\operatorname{var}\left(\mathbf{c}'\begin{pmatrix} \bar{X}_{1\cdot} \\ \vdots \\ \bar{X}_{I\cdot} \end{pmatrix}\right)}}{\sigma} \sqrt{(I-1)F_{\alpha, I-1, I(J-1)}\mathrm{MS_e}},\right.$$

$$\left.\mathbf{c}'\begin{pmatrix} \bar{X}_{1\cdot} \\ \vdots \\ \bar{X}_{I\cdot} \end{pmatrix} + \frac{\sqrt{\operatorname{var}\left(\mathbf{c}'\begin{pmatrix} \bar{X}_{1\cdot} \\ \vdots \\ \bar{X}_{I\cdot} \end{pmatrix}\right)}}{\sigma} \sqrt{(I-1)F_{\alpha, I-1, I(J-1)}\mathrm{MS_e}}\right).$$

Example (*Two-way ANOVA without interaction or randomized block design*)

$$X_{ijk} = \mu + \alpha_i + \beta_j + e_{ijk}, \qquad i = 1, \ldots, I, \quad j = 1, \ldots, J, \quad k = 1, \ldots, K;$$

$$\sum_i \alpha_i = \sum_j \beta_j = 0.$$

(a) Find $(1-\alpha)100\%$ simultaneous confidence intervals for the α_i main effects.

(i) T *method* The α_i's are contrasts among $E(\bar{X}_{i\cdot\cdot}) = \mu + \alpha_i$ (since $\bar{X}_{i\cdot\cdot} - (1/I)\sum_{i'=1}^{I} \bar{X}_{i'\cdot\cdot}$ has Σ coefficients $= 0$);

$$\operatorname{var}(\bar{X}_{i\cdot\cdot}) = \sigma^2/JK, \qquad \operatorname{cov}(\bar{X}_{i\cdot\cdot}, \bar{X}_{i'\cdot\cdot}) = 0, \qquad i \neq i';$$

$$\alpha_i \text{ has the interval} \quad \bar{X}_{i\cdot\cdot} - \bar{X}_{\cdot\cdot\cdot} \pm \frac{1}{\sqrt{JK}} q_{\alpha,I,IJK-I-J+1}\sqrt{\mathrm{MS_e}}\left(1 - \frac{1}{I}\right)$$

since

$$\tfrac{1}{2}\sum |c_i| = \tfrac{1}{2}\left((I-1)\frac{1}{I} + 1\left(1 - \frac{1}{I}\right)\right) = 1 - \frac{1}{I}.$$

(ii) S *method* We work in the $(I-1)$-dimensional subspace generated by $\bar{X}_{i\cdot\cdot} - \bar{X}_{\cdot\cdot\cdot}$, $i = 1, \ldots, I-1$;

$$\operatorname{var}(\bar{X}_{i\cdot\cdot} - \bar{X}_{\cdot\cdot\cdot}) = \operatorname{var}\left(\left(1 - \frac{1}{I}\right)\bar{X}_{i\cdot\cdot}\right) + \sum_{i' \neq i} \operatorname{var}\left(-\frac{1}{I}\bar{X}_{i'\cdot\cdot}\right)$$

$$= \left(1 - \frac{1}{I}\right)^2 \frac{\sigma^2}{JK} + (I-1)\left(\frac{1}{I}\right)^2 \frac{\sigma^2}{JK} = \left(1 - \frac{1}{I}\right)\frac{\sigma^2}{JK};$$

$$\alpha_i \text{ is in} \quad \bar{X}_{i\cdot\cdot} - \bar{X}_{\cdot\cdot\cdot} \pm \sqrt{\left(1 - \frac{1}{I}\right)\frac{1}{JK}}\sqrt{(I-1)F_{\alpha,I-1,IJK-I-J+1}\mathrm{MS_e}}.$$

(b) Use the S method to find $(1-\alpha)100\%$ simultaneous confidence intervals for μ, α_i, β_j for all i, j. In this case, we work in an $(I+J-1)$-dimensional space generated by $\bar{X}_{\cdot\cdot\cdot}, \bar{X}_{i\cdot\cdot} - \bar{X}_{\cdot\cdot\cdot}, \bar{X}_{\cdot j\cdot} - \bar{X}_{\cdot\cdot\cdot}$. The simultaneous intervals are

$$\bar{X}_{\cdot\cdot\cdot} \pm \sqrt{\frac{1}{IJK}}\sqrt{(I+J-1)F_{\alpha,I+J-1,IJK-J-I+1}\mathrm{MS_e}} \qquad \text{for } \mu,$$

$$\bar{X}_{i\cdot\cdot} - \bar{X}_{\cdot\cdot\cdot} \pm \sqrt{\left(1 - \frac{1}{I}\right)\frac{1}{JK}}\sqrt{(I+J-1)F_{\alpha,I+J-1,IJK-I-J+1}\mathrm{MS_e}} \qquad \text{for } \alpha_i,$$

$$\bar{X}_{\cdot j\cdot} - \bar{X}_{\cdot\cdot\cdot} \pm \sqrt{\left(1 - \frac{1}{J}\right)\frac{1}{IK}}\sqrt{(I+J-1)F_{\alpha,I+J-1,IJK-I-J+1}\mathrm{MS_e}} \qquad \text{for } \beta_j.$$

Example 8.2 Consider again Example 4.1 dealing with drug dosage schemes in mixed anxiety/depressive states. The authors compare the different treatments by using the S method. Their data are presented in Tables 8.2 and 8.3. From this (Table 8.2), the authors conclude that one tablet three times a day (1 tablet t.d.s.) or one tablet at night is a regime preferable to one tablet in the morning.

TABLE 8.2

Pairwise Difference between Treatment Groups: Scheffé Multiple Comparison Test on Physician's Clinical Rating Scale Group Improvements

Period	Treatment pair	Mean difference/ standard error	Significance[a]
Day 0 to Day 7	1 tablet t.d.s./1 tablet morning	2.57	$p < .05$
	1 tablet t.d.s./1 tablet night	−.54	NS
	1 tablet morning/1 tablet night	−3.08	$p < .05$
Day 0 to Day 28	1 tablet t.d.s./1 tablet morning	3.93	$p < .001$
	1 tablet t.d.s./1 tablet night	.92	NS
	1 tablet morning/1 tablet night	−3.04	$p < .05$

[a] NS = not significant.

TABLE 8.3

Pairwise Difference between Treatment Groups: Scheffé Multiple Comparison Test on Patient's Visual Analogue Scale Group Improvements

Period	Treatment pair	Mean difference/ standard error	Significance[a]
Day 0 to Day 7	1 tablet t.d.s./1 tablet morning	1.56	NS
	1 tablet t.d.s./1 tablet night	−.29	NS
	1 tablet morning/1 tablet night	−1.84	NS
Day 0 to Day 28	1 tablet t.d.s./1 tablet morning	2.39	NS
	1 tablet t.d.s./1 tablet night	.80	NS
	1 tablet morning/1 tablet night	−1.60	NS

[a] NS = not significant.

Example 8.3 Example 6.2 deals with commercials and violence. For the overall effectiveness data as presented in Tables 6.4 and 6.5, we construct 95% S- and T-method confidence intervals for the film main effects. This will be an extension of (a) above with a different degree of freedom associated with MS_e.

Now $I = 3$, $J = 2$, $K = 4$. Further, the means for the eight groups seeing each film are:

Wild One	$\bar{X}_{1..} = 13.13,$	
Attica	$\bar{X}_{2..} = 12.17,$	$\bar{X}_{...} = 11.20.$
Mouse That Roared	$\bar{X}_{3..} = 8.30,$	

First, find the T-method confidence intervals. Using the tables in Appendix 2, $q_{.05,3,18} = 3.61$, $MS_e = 1.97$. The confidence intervals are

$$\bar{X}_{i..} - \bar{X}_{...} \pm \frac{1}{\sqrt{JK}} q_{\alpha,I,IJ(K-1)} \sqrt{MS_e\left(1 - \frac{1}{I}\right)}.$$

Thus, the T-method 95% simultaneous confidence intervals are

Wild One	(.74, 3.2),
Attica	(.22, 2.16),
Mouse That Roared	(−4.09, −1.71).

Turning to the S method, we have $F_{.05,2,18} = 3.55$. The intervals are

$$\bar{X}_{i\cdot\cdot} - \bar{X}_{\cdot\cdot\cdot} \pm \sqrt{\left(1 - \frac{1}{I}\right)\frac{1}{JK}}\sqrt{(I-1)F_{\alpha,I-1,IJ(K-1)}\mathrm{MS_e}}.$$

Thus the S-method 95% simultaneous confidence intervals are

Wild One	(.85, 3.01),
Attica	(.33, 2.05),
Mouse That Roared	(−3.98, −1.82).

The main effects are subject to careful interpretation since the interaction terms are significantly nonzero in this example. The main effects for films indicate subjects were more agressive following the aggressive films than following the nonaggressive one. (Note that the *Mouse That Roared* confidence interval does not intersect the other two confidence intervals.)

Subjects were also more aggressive in the commercial conditions than in the no-commercial conditions. These main effects were also qualified by the significant interaction. The commercials significantly increased aggression in the two aggressive film conditions ($p < .01$ studentized range statistic) but not in the nonaggressive film condition. Within the commercial condition, there was significantly more aggression in the aggressive film conditions than in the nonaggressive movie cell. Finally, within the no-commercial cells, the only comparison to reach significance ($p < .05$) was that between *Wild One* and *Mouse That Roared.*

SUMMARY

This chapter considers ANOVA situations in which one desires several confidence intervals that all hold simultaneously. Equivalently, one may simultaneously make several tests or compare several quantities.

The generalization of a confidence interval is a confidence set. The analysis of variance lends itself naturally to ellipsoidal confidence sets.

THEOREM Let $\mathbf{X}^{n\times 1} \sim N(\mathbf{D}\boldsymbol{\beta}, \sigma^2\mathbf{I})$, $\mathbf{D}$ of rank p, and $\hat{\boldsymbol{\beta}}$ be the least squares estimator of $\boldsymbol{\beta}^{p\times 1}$. Then a $100(1-\alpha)\%$ confidence set for $\boldsymbol{\beta}$ is $\{\tilde{\boldsymbol{\beta}}:(\hat{\boldsymbol{\beta}} - \tilde{\boldsymbol{\beta}})'\mathbf{D}'\mathbf{D}(\hat{\boldsymbol{\beta}} - \tilde{\boldsymbol{\beta}})/(p\mathrm{MS_e}) \le F_{\alpha,p,n-p}\}$. Three approaches to simultaneous confidence intervals were considered.

(a) *Bonferroni's inequality* This is a general purpose approach that holds in many situations.

LEMMA Let $I_i(\mathbf{X})$ be a $100(1-\alpha_i)\%$ confidence interval for the parameter γ_i, $i = 1, \ldots, k$. The probability that all k parameters are simultaneously in their confidence intervals is greater than or equal to $1 - \sum_{i=1}^{k} \alpha_i$; that is,

$$P\left(\bigcap_{i=1}^{k} (\gamma_i \in I_i(\mathbf{X}))\right) \geq 1 - \sum_{i=1}^{k} \alpha_i.$$

(b) *The* S *method or Scheffé's method* This approach uses planes of support in the preceding theorem. It allows simultaneous confidence intervals for a subspace of linear combinations of a multinormal random vector.

THEOREM Let $\tilde{\gamma}^{k \times 1} \sim N(\gamma, \sigma^2 \mathbf{C})$, where $\sigma^2 > 0$ and $\mathbf{C}$ is positive definite. Independently of $\tilde{\gamma}$, let $ms^2/\sigma^2 \sim \chi_m{}^2$. Then with confidence $1 - \alpha$ *for all* $\mathbf{c}$,

$$\mathbf{c}'\gamma \text{ is in } \quad (\mathbf{c}'\tilde{\gamma} - \sqrt{\mathbf{c}'\mathbf{C}\mathbf{c}}\sqrt{kF_{\alpha,k,m}}s, \mathbf{c}'\tilde{\gamma} + \sqrt{\mathbf{c}'\mathbf{C}\mathbf{c}}\sqrt{kF_{\alpha,k,m}}s).$$

$F_{\alpha,k,m}$ is the α critical value of the F-distribution with k and m df.

(c) *The* T *method or Tukey's method* This approach works only in certain balanced situations.

DEFINITION A *contrast* of parameters

$$\boldsymbol{\beta} = \begin{pmatrix} \beta_1 \\ \beta_2 \\ \vdots \\ \beta_p \end{pmatrix}$$

is a sum $\sum_{i=1}^{p} c_i \beta_i$, where $\sum_{i=1}^{p} c_i = 0$. A *simple contrast* has one $c_i = 1$, one $c_{i'} = -1$, and the remainder equal to zero.

The T method is usually preferable for simple contrasts and the S method for more complex situations.

THEOREM Let (a) $Z_1, \ldots, Z_k$ be $N(\mu_1, a\sigma^2), \ldots, N(\mu_k, a\sigma^2)$ respectively, and the Z_i's are multinormally distributed. Let $\text{cov}(Z_i, Z_j) = b\sigma^2$, $i \neq j$, where $b \leq 0$.

(b) Let s^2 be independent of $Z_1, \ldots, Z_k$, and $ms^2/\sigma^2 \sim \chi^2$ with m df.

Then the probability is $1 - \alpha$ that for all $\mathbf{c}' = (c_1, \ldots, c_k)$ with $\mathbf{c}'\mathbf{1} = 0$,

$$\mathbf{c}'\boldsymbol{\mu} \text{ is in } \left(\mathbf{c}'\mathbf{Z} - \sqrt{a-b}\, q_{\alpha,k,m} s \tfrac{1}{2} \sum_{i=1}^{k} |c_i|, \mathbf{c}'\mathbf{Z} + \sqrt{a-b}\, q_{\alpha,k,m} s \tfrac{1}{2} \sum_{i=1}^{k} |c_i|\right).$$

where $\mathbf{Z}' = (Z_1, \ldots, Z_k)$ and $\boldsymbol{\mu}' = (\mu_1, \ldots, \mu_k)$.

PROBLEMS

8.1 In Theorem 8.1 with $p = 2$, show that the confidence set is an ellipse.

8.2 In the proof of Theorem 8.2, show that $\|\mathbf{D}^{1/2}\mathbf{Pc}\| = \sqrt{\mathbf{c'Cc}}$.

8.3 Prove Corollary 8.3. [*Hint*: What is $\mathbf{D(D'D)^{-1}D'}$?]

8.4 Suppose there are two factors A and B taking two levels each with no interaction; that is, if A is at level i ($i = 1$ or 2) and B is at level j ($j = 1$ or 2), then the expected value of an observation is $\mu + \alpha_i + \beta_j$ (where $\alpha_1 + \alpha_2 = 0$, $\beta_1 + \beta_2 = 0$). The errors of observation are i.i.d. $N(0, \sigma^2)$. Five observations are taken (two at the 1–1 level) and give the values shown in Table 8.4.

(a) Find the least squares estimates for μ, α_1, and β_1. [*Hint*: Write the design matrix in terms of these three parameters.]

(b) If you are interested only in α_1, find a 90% confidence interval for α_1.

(c) Find the simultaneous 90% S-confidence intervals for μ, α_1, and β_1.

(d)* Use the Bonferroni approach to find simultaneous 90% confidence intervals for μ, α_1, and β_1. [*Hint*: Each individual interval should have $(100 - (10/3))\%$ confidence. You need to interpolate from the F tables given in Appendix 2.]

TABLE 8.4

	Factor A	
Factor B	Level 1	Level 2
Level 1	5.27	4.75
	5.64	
Level 2	6.08	5.21

8.5 Show that the studentized range $Q_{k,m}$ has a distribution that does not depend on μ or σ^2.

8.6 Let S be a p-dimensional subspace of R^n. Show that

$$\max_{\mathbf{a} \in S, \mathbf{a} \neq \mathbf{0}} \left(\frac{\mathbf{a'y}}{\|\mathbf{a}\|}\right)^2 = \|\mathbf{P}_S \mathbf{y}\|^2,$$

where $\mathbf{P}_S$ is the projection operator into S.

8.7 Prove by using Problem 8.6 that (Corollary 8.4 restated):

Let $\mathbf{X} \sim N(\boldsymbol{\mu}, \sigma^2\mathbf{I})$. Let $(m\,\mathrm{MS_e}/\sigma^2) \sim \chi_m{}^2$ independently of $\mathbf{P}_S\mathbf{X}$. Then

$$P(|\mathbf{a'X} - \mathbf{a'}\boldsymbol{\mu}| \leq \sqrt{pF_{\alpha,p,m}\mathrm{MS_e}}\|\mathbf{a}\| \text{ for all } \mathbf{a} \in S) = 1 - \alpha.$$

[*Hint*: Show that the event being considered is the same as $\|\mathbf{P}_S(\mathbf{X} - \boldsymbol{\mu})\|^2/p \leq F_{\alpha,p,m}\mathrm{MS_e}$.]

8.8 Suppose we have a 4×4 Latin Square design.

(a) Set up T- and S-confidence intervals for the α_1, α_2, α_3, and α_4.

(b) Can you set up simultaneous T intervals for the α_i's and the β_j's? If so, do it; if not, explain why not.

(c) Repeat part (b) with "S" replacing "T."

8.9 For the Latin Square data of Example 7.2, find:

(a) S- and T-95% simultaneous confidence intervals for the treatment main effects.

(b) S- and T-95% simultaneous confidence intervals for the simple contrasts between treatment means. Are the M, P, and/or S treatments superior to the control treatment O?

8.10 For the Latin Square data of Example 7.3, find simultaneous 95% confidence intervals for the treatment means *and* the contrasts between means. By examining the intervals, what can you conclude about the differences (if any) between treatments?

8.11 Example 8.1 of the procainamide study uses Bonferroni's inequality to take into account the 15 tests being made. From Table 8.1, set up a 95% confidence interval for the difference in resting heart rate as if 9 other simultaneous confidence intervals were being constructed. Use Bonferroni's inequality. [*Hint*: Use $\alpha = .05$; from Table 8.1 you see that $5.053 = t = \sqrt{23}(87 - 73)/s$ giving the value of s.]

8.12 The study of Examples 4.1 and 8.2 gave Table 8.5 for the improvement on the physician's clinical rating scale.

(a) Using this table and the ANOVA table of Example 4.1 (Table 4.1), find three simultaneous 95% confidence intervals for the mean differences between the groups in the improvement in the physician's clinical rating scale.

(b) In Table 8.2 verify the value 2.57. How does one find the corresponding p value?

TABLE 8.5

Physician's Clinical Rating Scale: Mean Within-Group Improvements and Significance Values by Paired t-Test

Treatment group	Period	No. patients	Mean (±SEM) decrease in score, i.e., improvement[a]
1 tablet t.d.s.	Day 0 to Day 7	77	5.0 ± .40
	Day 0 to Day 28	72	10.69 ± .61
1 tablet in morning	Day 0 to Day 7	71	3.51 + .42
	Day 0 to Day 28	69	7.26 ± .62
1 tablet at night	Day 0 to Day 7	75	5.31 ± .41
	Day 0 to Day 28	73	9.90 ± .61

[a] Improvements within each group significantly different from zero at $p < .001$.

8.13 Using the results of Problem 4.9, set up simultaneous 90% confidence intervals for the simple contrasts between mandibular fluoride in the five groups. Which means differ (at a 10% simultaneous significance level)?

(a) Use the T method.

(b) Use the S method.

8.14 Using Tukey's studentized range procedure (the T method), verify that the superscripts in Table 4.6 (Problem 4.10) do indeed give means that differ at the .05 significance level (i.e., find simultaneous 95% T-method confidence intervals for the simple contrasts and examine which intervals do not contain zero).

8.15 Consider again Example 6.2 on commercials and violence. The numbers in parentheses at the bottom of Table 6.5 are the estimates of σ^2, MS_e, for the analyses.

From Tables 6.4 and 6.5, find:

(a) simultaneous S-method 90% confidence intervals for the six cell means on the "should hire" data;

(b) simultaneous S-method 95% confidence intervals for the interaction terms γ_{ij} for the "should hire" data;

(c) 90% T-method confidence intervals for the simple contrasts between the film ratings for the "should hire" data;

(d) repeat part (c) but use the S method for the "should hire" data.

8.16 Consider again the data of Example 5.1 which examined the plasma-free methionine level in rats fed heated soybean protein.

(a) Find simultaneous 99% S- and T-confidence intervals for the heat treatment simple contrasts.

(b) Find simultaneous 90% confidence intervals for all the cell means assuming that the model

$$\mu_{ij} = \mu + \alpha_i + \beta_j$$

is correct. (S method, of course. Why?)

(c) Find simultaneous 95% confidence intervals for the α_i and β_j.

9 ORTHOGONAL AND NONORTHOGONAL DESIGNS, EFFICIENCY

One of the main themes of this text has been that of projecting the data vector onto orthogonal subspaces. For multinormal observations with a covariance matrix of the form $\sigma^2\mathbf{I}$, the projections of the data vector onto orthogonal subspaces are independently distributed multinormal random vectors. The methods resulting from this fact have been aesthetically pleasing—the analysis of variance being primarily a study in decomposing a vector into its orthogonal components and using the Pythagorean theorem. The problems we have had resulted mainly from recording data in the "wrong" coordinate system for the decomposition at hand. All of this, however, would be poor statistics if these decompositions were a wasteful (in terms of the number of observations needed) method of collecting and analyzing the data. In this chapter we shall see why, as a rough rule of thumb, "the more orthogonal a design an experiment has, the more efficient the design." The rule of thumb, while not invariably true, has quite a bit of truth to it for reasons discussed in this chapter. It is pleasing that an aesthetic way of proceeding is also efficient—not an invariable fact of life in human affairs.

As a beginning, consider an overly simplified case. Suppose $\mathbf{X} \sim N(\alpha\mathbf{v} + \beta\mathbf{w}, \sigma^2\mathbf{I})$, where α, β, and σ^2 are unknown parameters; $\mathbf{v}$ and $\mathbf{w}$ are unit length vectors chosen by the experimenter and may have any orientation in R^n. Without loss of generality, we will suppose that $\mathbf{v}$ and $\mathbf{w}$ lie in the plane of the page in order to facilitate the drawings. The other $n - 2$ dimensions of the error subspace are orthogonal to the page and are left to your imagination. The only real choice we have is to choose the angle θ between $\mathbf{v}$ and $\mathbf{w}$ (Figure 9.1). The design matrix is $\mathbf{D} = (\mathbf{v}\ \mathbf{w})$. It is easily shown that the

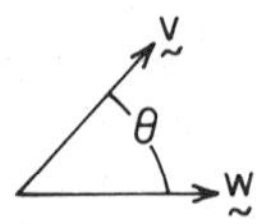

FIGURE 9.1

distribution of the least squares estimates $\hat{\alpha}$ and $\hat{\beta}$ of α and β can be written

$$\begin{pmatrix}\hat{\alpha}\\ \hat{\beta}\end{pmatrix} \sim N\left(\begin{pmatrix}\alpha\\ \beta\end{pmatrix}, \sigma^2(\mathbf{D}'\mathbf{D})^{-1}\right) = N\left(\begin{pmatrix}\alpha\\ \beta\end{pmatrix}, \frac{\sigma^2}{\sin^2\theta}\begin{pmatrix}1 & -\cos\theta\\ -\cos\theta & 1\end{pmatrix}\right). \tag{1}$$

The variances of $\hat{\alpha}$ and $\hat{\beta}$ are $\sigma^2/\sin^2\theta$ which are minimized by taking $\theta = 90°$ or $270°$; that is, by choosing $\mathbf{v}$ and $\mathbf{w}$ to be orthogonal. As a necessary by-product in these cases, $\text{cov}(\hat{\alpha}, \hat{\beta}) = 0$, so that these estimates are also independent.

Now consider the least squares estimates:

$$\hat{\alpha} = \frac{\mathbf{v}'\mathbf{X}}{\sin^2\theta} - \frac{\mathbf{w}'\mathbf{X}\cos\theta}{\sin^2\theta}, \qquad \hat{\beta} = \frac{\mathbf{w}'\mathbf{X}}{\sin^2\theta} - \frac{\mathbf{v}'\mathbf{X}\cos\theta}{\sin^2\theta}.$$

Let us see geometrically from where these estimates come (Figure 9.2). We cannot estimate α directly by projecting the data vector onto the $\mathbf{v}$ direction since the $\beta\mathbf{w}$ "confounds" the α estimate, that is, $E(\mathbf{v}'\mathbf{X}) = \mathbf{v}'(\alpha\mathbf{v} + \beta\mathbf{w}) = \alpha + \beta\cos\theta$, and unless $\cos\theta = 0$, this is not an unbiased estimate of α. To get around this, we take the component of $\alpha\mathbf{v}$ that is orthogonal to $\mathbf{w}$ and project the data vector onto this direction. Let $\mathbf{u}$ be a unit vector in the $\mathbf{vw}$ plane that is orthogonal to $\mathbf{w}$ and forms an angle less than 90° with $\mathbf{v}$. The projection of $\alpha\mathbf{v}$ onto the $\mathbf{u}$ direction is $\alpha\mathbf{u}(\mathbf{u}'\mathbf{v}) = \alpha\mathbf{u}\sin\theta$. Thus, since $E(\mathbf{u}'\mathbf{X}) = \mathbf{u}'(\alpha\mathbf{v} + \beta\mathbf{w}) = \alpha\sin\theta$, we can estimate α by $\mathbf{u}'\mathbf{X}/\sin\theta$. Note, of course, that $\alpha\sin\theta\,\mathbf{u}$ has a smaller length than $\alpha\mathbf{v}$, so that the quantity α is more difficult to estimate; $\text{var}(\mathbf{u}'\mathbf{X}/\sin\theta) = \sigma^2/\sin^2\theta$. It is easy to verify that

$$\mathbf{u} = \frac{\mathbf{v}}{\sin\theta} - \frac{\cos\theta}{\sin\theta}\mathbf{w} \tag{2}$$

and to use this relationship to show $\hat{\alpha} = \mathbf{u}'\mathbf{X}/\sin\theta$.

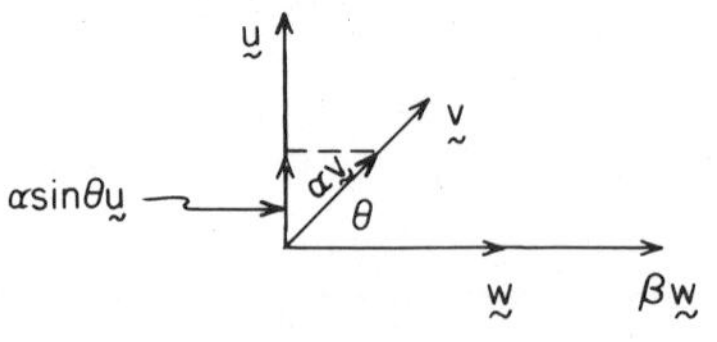

FIGURE 9.2

Note that if θ is small, the variance of the estimates for α and β becomes extremely large. This problem is often referred to as the "problem of multicollinearity." If the columns of the design matrix associated with the two parameters are "nearly" parallel, that is, "nearly" collinear, then it is difficult to estimate both parameters.

The term "confounding" or "confounded" is used in statistics when it is difficult (statistically) to separate out the effects of two (or more) factors. Thus, in this example, if θ is small, α and β are confounded. Quantities are completely confounded if it is impossible to estimate them, because different combinations of values give the same observational distribution for $\mathbf{X}$. In this example, if $\theta = 0°$, then α and β are completely confounded (see Problem 9.4).

Suppose, however, we are not interested in separating out the α and β coefficients but only desire to estimate the *combined* effect of α and β. In this case, we estimate the effect by the projection of $\mathbf{X}$ onto the $\mathbf{vw}$ plane. The estimate (projection) will be the same regardless of which pair of vectors $\mathbf{v}$ and $\mathbf{w}$ in the plane we select. From these considerations we should revise our rule of thumb. "For efficiency in a design, we want orthogonality (or as nearly as possible) between subsets of parameters whose component effects we wish to separate out. Combinations of variables that we need not "untangle" are not required to be at all orthogonal (as long as the columns of the design matrix span an appropriate space)." Rephrasing our rule of thumb, we have: *Effects that we are interested in examining separately should be designed or observed as orthogonally as possible.* This sentence abuses the terminology, but the idea is clear.

Consider the reverse side of the coin. Suppose you design and analyze an experiment to investigate the factors. The design confounds the two factors severely. At the end of the analysis, the following dialogue ensues:

You: "As you expected, beyond any shadow of a doubt the two factors have a huge effect ($p < .00001$) on the outcome."

Experimenter: "Great! Which factor is responsible? Or are both clearly contributing?"

You: "Uh . . . Well . . . I don't quite know how to tell you this—but we can't show that either factor has an effect."

(Time passes with confused interchange.)

Experimenter (leaving in disgust): "Wait until I find the fellow who told me to consult a statistician *before* running the experiment!"

The point here is that $(\mathbf{D'D})^{-1}$ of a proposed design may be examined before an experiment (and, if the experimenter has enough control, adjusted if need be to allow accurate estimation of important parameters.)

Before moving on to the next example, consider the test to use for testing $\alpha = 0$ versus $\alpha \neq 0$. We would like the sum of squares associated with the **u** direction. Often, the easiest way to find the SS_α is by the Pythagorean theorem. Let $SS_{\alpha,\beta}$ be the SS from projecting onto the **vw** plane and $SS_{\mathbf{w}}$ the SS from projecting onto the **w** direction; then the SS_α is

$$SS_\alpha = SS_{\alpha,\beta} - SS_{\mathbf{w}},$$

and the appropriate F-test (with MS_e the error mean square) is to use

$$F = \frac{SS_{\alpha,\beta} - SS_{\mathbf{w}}}{MS_e} \sim F_{1,n-2}\left(\frac{|\alpha \sin\theta|}{\sigma}\right). \tag{3}$$

Note the low statistical power (i.e., small noncentrality parameter) if θ is such that $\sin\theta$ is small.

The directions of the column vectors are determined by the pattern of observations. As an example of how the geometric intuition relates to the allocation of observations in an experimental design, consider the following situation. There are two factors A and B, each taking two levels. The model is

$$X_{ijk} = \mu + \alpha_i + \beta_j + e_{ijk},$$

where $i = 1$ or 2 denotes the level of factor A, $j = 1$ or 2 the level of factor B, $\alpha_1 + \alpha_2 = 0 = \beta_1 + \beta_2$ and $k(i,j) = 1, 2, \ldots, n_{ij}$. The e_{ijk} are i.i.d. $N(0, \sigma^2)$ random variables. To design the experiment, we may allocate $N = 4n$ observations, so that $n_{11} + n_{12} + n_{21} + n_{22} = N$. Rewriting the expected values of the observations in terms of $\alpha \equiv \alpha_1$ and $\beta \equiv \beta_1$, we have the presentation shown in Table 9.1. Forming the observation vector $\mathbf{X}$ by lexicographically ordering the X_{ijk}, we obtain

$$\mathbf{D} = \begin{pmatrix} 1 & 1 & 1 \\ \vdots & \vdots & \vdots \\ 1 & 1 & 1 \\ 1 & 1 & -1 \\ \vdots & \vdots & \vdots \\ 1 & 1 & -1 \\ 1 & -1 & 1 \\ \vdots & \vdots & \vdots \\ 1 & -1 & 1 \\ 1 & -1 & -1 \\ \vdots & \vdots & \vdots \\ 1 & -1 & -1 \end{pmatrix} \begin{matrix} \\ n_{11} \text{ rows} \\ \\ \\ n_{12} \text{ rows} \\ \\ \\ n_{21} \text{ rows} \\ \\ \\ n_{22} \text{ rows} \\ \\ \end{matrix} \qquad \text{and} \qquad \boldsymbol{\beta} = \begin{pmatrix} \mu \\ \alpha \\ \beta \end{pmatrix}.$$

$$\begin{matrix} \uparrow & \uparrow & \uparrow \\ \mathbf{v}_\mu & \mathbf{v}_\alpha & \mathbf{v}_\beta \end{matrix}$$

TABLE 9.1

	Factor B	
Factor A	Level 1	Level 2
Level 1	$\mu + \alpha + \beta$ n_{11} observations	$\mu + \alpha - \beta$ n_{12} observations
Level 2	$\mu - \alpha + \beta$ n_{21} observations	$\mu - \alpha - \beta$ n_{22} observations

Let θ_1 be the angle between $\mathbf{v}_\mu$ and $\mathbf{v}_\alpha$, θ_2 the angle between $\mathbf{v}_\mu$ and $\mathbf{v}_\beta$, and θ_3 the angle between $\mathbf{v}_\alpha$ and $\mathbf{v}_\beta$. Direct calculation then gives

$$\mathbf{D}'\mathbf{D} = \begin{pmatrix} N & (n_{11} + n_{12} - n_{21} - n_{12}) & (n_{11} - n_{12} + n_{21} - n_{22}) \\ & N & (n_{11} - n_{12} - n_{21} + n_{22}) \\ \text{Symmetric} & & N \end{pmatrix}$$

$$= \begin{pmatrix} \|\mathbf{v}_\mu\|^2 & \|\mathbf{v}_\mu\|\,\|\mathbf{v}_\alpha\|\cos\theta_1 & \|\mathbf{v}_\mu\|\,\|\mathbf{v}_\beta\|\cos\theta_2 \\ & \|\mathbf{v}_\alpha\|^2 & \|\mathbf{v}_\alpha\|\,\|\mathbf{v}_\beta\|\cos\theta_3 \\ \text{Symmetric} & & \|\mathbf{v}_\beta\|^2 \end{pmatrix}$$

since $\|\mathbf{v}_\mu\|^2 = \|\mathbf{v}_\alpha\|^2 = \|\mathbf{v}_\beta\|^2 = N$ and

$$\cos\theta_1 = (n_{11} + n_{12} - n_{21} - n_{22})/N,$$
$$\cos\theta_2 = (n_{11} - n_{12} + n_{21} - n_{22})/N,$$
$$\cos\theta_3 = (n_{11} - n_{12} - n_{21} + n_{22})/N.$$

Further,

$$|\mathbf{D}'\mathbf{D}| = \left| N \begin{pmatrix} 1 & \cos\theta_1 & \cos\theta_2 \\ \cos\theta_1 & 1 & \cos\theta_3 \\ \cos\theta_2 & \cos\theta_3 & 1 \end{pmatrix} \right|$$
$$= N^3(1 + 2\cos\theta_1\cos\theta_2\cos\theta_3 - \cos^2\theta_1 - \cos^2\theta_2 - \cos^2\theta_3).$$

The covariance matrix of the least squares estimates

$$\begin{pmatrix} \hat{\mu} \\ \hat{\alpha} \\ \hat{\beta} \end{pmatrix}$$

is

$$\frac{N^2\sigma^2}{|\mathbf{D}'\mathbf{D}|} \begin{pmatrix} \sin^2\theta_3 & \cos\theta_2\cos\theta_3 - \cos\theta_1 & \cos\theta_1\cos\theta_3 - \cos\theta_2 \\ & \sin^2\theta_2 & \cos\theta_1\cos\theta_2 - \cos\theta_3 \\ \text{Symmetric} & & \sin^2\theta_1 \end{pmatrix}.$$

How might we minimize the variance of, say, $\hat{\alpha}$?

$$\begin{aligned}\operatorname{var}(\hat{\alpha}) &= \frac{\sin^2\theta_2\sigma^2}{N|\sin^2\theta_2 + 2\cos\theta_1\cos\theta_2\cos\theta_3 - \cos^2\theta_1 - \cos^2\theta_3|} \\ &\geq \frac{\sin^2\theta_2\sigma^2}{N|\sin^2\theta_2 + 2|\cos\theta_1||\cos\theta_3| - \cos^2\theta_1 - \cos^2\theta_3|} \\ &= \frac{\sin^2\theta_2\sigma^2}{N|\sin^2\theta_2 - (|\cos\theta_3| - |\cos\theta_1|)^2|} \geq \frac{\sigma^2}{N}.\end{aligned}$$

Now $\operatorname{var}(\hat{\alpha})$ is equal to σ^2/N when $\cos\theta_1 = \cos\theta_3 = 0$ and $|\sin\theta_2| > 0$. Similar arguments show that we may simultaneously minimize $\operatorname{var}(\hat{\mu})$, $\operatorname{var}(\hat{\alpha})$, and $\operatorname{var}(\hat{\beta})$ by choosing $\cos\theta_1 = \cos\theta_2 = \cos\theta_3 = 0$, each estimate having variance σ^2/N, i.e., $n_{11} = n_{12} = n_{21} = n_{22}$.

Thus, in this situation the most accurate estimates occur when the observations are distributed evenly over the four possibilities. In other more complex situations, it may occur that accuracy in estimating one parameter may decrease the accuracy in estimating other parameters.

The result we have arrived at is astounding. Consider: if we had N i.i.d. $N(\alpha, \sigma^2)$ random variables, the estimate of α would be $\bar{X}$ with a variance of σ^2/N. In this example, however, *three* quantities are estimated, all with the same variance one would have if there were only one parameter to estimate! (There is a slight price to pay for this magic as there are only $N - 3$ dimensions or degrees of freedom to estimate σ^2 instead of $N - 1$.) The possible efficiency also depends on the assumed model being (approximately) correct. This illustrates part of what Sir R. A. Fisher meant when he said:

> No aphorism is more frequently repeated with field trials, than that we must ask Nature few questions, or ideally, one question at a time. The writer is convinced that this view is wholly mistaken. Nature he suggests, will best respond to a logical and carefully thought out questionnaire; indeed, if we ask her a single question, she will often refuse to answer until some other topic has been discussed.

Another aspect of this problem is typical of good experimental designs. The efficient orthogonal designs are mirrored by orderly symmetric assignment of observations to the possibilities. The subject of efficient design of experiments has been greatly developed, and the example discussed here only remotely indicates the possibilities.

Consider the general case of our usual linear model. The least squares estimates, their variances, and covariances depend on both the length and orientation of the columns of the design matrix.

THEOREM 9.1 Let $E(\mathbf{X}) = \mathbf{D}\boldsymbol{\beta}$, where the design matrix $\mathbf{D}$ has columns $\mathbf{d}_1, \ldots, \mathbf{d}_p$ and full column rank p, the dimension of $\boldsymbol{\beta}$, our vector of unknown parameters. Also assume that the covariance of $\mathbf{X}$ is $\sigma^2\mathbf{I}$. Let $\mathbf{P}_{(-i)}$ denote the projection onto the subspace spanned by all the columns of the design matrix *except* the ith column. Let $\xi_i = \mathbf{d}_i - \mathbf{P}_{(-i)}\mathbf{d}_i$; that is, the component of $\mathbf{d}_i$ which is orthogonal to the subspace spanned by all the columns of the design matrix except the ith column. Then if $\hat{\beta}_i$ is the least squares estimate of β_i, we have

$$\hat{\beta}_i = \frac{\xi_i'\mathbf{X}}{\|\xi_i\|^2}, \qquad \operatorname{var}(\hat{\beta}_i) = \frac{\sigma^2}{\|\xi_i\|^2}, \qquad \operatorname{cov}(\hat{\beta}_i, \hat{\beta}_j) = \frac{\sigma^2\xi_i'\xi_j}{\|\xi_i\|^2\|\xi_j\|^2}.$$

Proof Problem 9.7. □

Now the length and orientation of the columns of the design matrix determine the length and orientation of the ξ_i's which determine our least squares estimates and their properties. Since $\operatorname{var}(\hat{\beta}_i) = \sigma^2/\|\xi_i\|^2$ and $\|\xi_i\|^2 = \|\mathbf{d}_i\|^2 - \|\mathbf{P}_{(-i)}\mathbf{d}_i\|^2$, $\operatorname{var}(\hat{\beta}_i)$ is minimized when $\|\mathbf{P}_{(-i)}\mathbf{d}_i\|^2$ equals zero, that is, when $\mathbf{d}_i$ is orthogonal to the remaining columns of the design matrix; $\operatorname{var}(\hat{\beta}_i)$ is also inversely proportional to the squared length of its corresponding column vector, thus each column of the design matrix should be chosen to have as large a length as possible. (The usual way of increasing length is to increase the sample size.)

Suppose $\mathbf{d}_1$ and $\mathbf{d}_2$ lie in the plane of the page and the remaining columns of the design matrix and the error subspace are orthogonal to the page and left to your imagination (Figure 9.3). We note that the ξ's are of smaller length unless the $\mathbf{d}$'s are orthogonal. Also note that as θ gets small, so do $\|\xi_1\|$ and $\|\xi_2\|$, and the variances of $\hat{\beta}_1$ and $\hat{\beta}_2$ become extremely large.

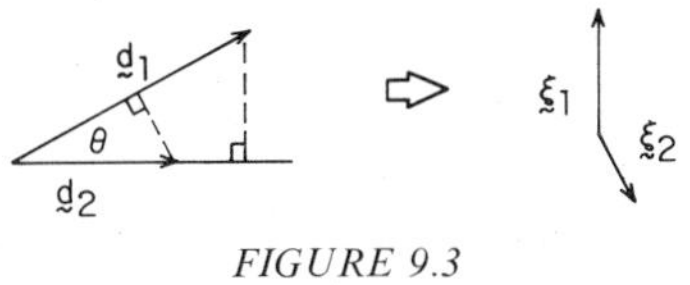

FIGURE 9.3

SUMMARY

Consider our usual linear model $E(\mathbf{X}) = \mathbf{D}\boldsymbol{\beta}$, $\operatorname{cov}(\mathbf{X}) = \sigma^2\mathbf{I}$. The length and orientation of the columns of the design matrix determine the length and orientation of the vectors $\xi_1, \ldots, \xi_p$, where ξ_i is the component of the ith column of the design matrix which is orthogonal to the subspace spanned by the remaining columns of the design matrix. The $\xi_1, \ldots, \xi_p$ determine our least squares estimates and their properties; namely, if $\hat{\beta}_i$ is the least squares estimate of β_i, then

$$\hat{\beta}_i = \frac{\xi_i'\mathbf{X}}{\|\xi_i\|^2}, \qquad \mathrm{var}(\hat{\beta}_i) = \frac{\sigma^2}{\|\xi_i\|^2}, \qquad \mathrm{cov}(\hat{\beta}_i, \hat{\beta}_j) = \frac{\sigma^2 \xi_i' \xi_j}{\|\xi_i\|^2 \|\xi_j\|^2}.$$

The vectors $\xi_1, \ldots, \xi_p$ are useful in geometrically understanding the "problem of multicollinearity" and in choosing a good experimental design.

PROBLEMS

9.1 Verify the covariance matrix in equation (1).

9.2 Show that $\mathrm{var}(\mathbf{u}'\mathbf{X}/\sin\theta) = \sigma^2/\sin^2\theta$.

9.3 Verify equation (2).

9.4 If $\theta = 0°$ or $180°$ in Figure 9.1, many of the formulas "blow up" as division by $\sin\theta$ takes place. Why should this be expected and not bother us?

9.5 In more detail than that given in the text, show that equation (3) holds.

9.6 Show that $\cos\theta_1 = \cos\theta_2 = \cos\theta_3 = 0$ implies $n_{11} = n_{12} = n_{21} = n_{22}$. More generally, show that if $\mathrm{var}(\hat{\mu}) = \mathrm{var}(\hat{\alpha}) = \mathrm{var}(\hat{\beta}) = \sigma^2/N$, then $n_{11} = n_{12} = n_{21} = n_{22}$ (see pages 120 to 122 for notation).

9.7 Prove Theorem 9.1. [*Hint*: Without loss of generality, assume $i = 1$. Reparameterize so that $E(\mathbf{X}) = \gamma_1\xi_1 + \gamma_2\mathbf{d}_2 + \cdots + \gamma_p\mathbf{d}_p$; that is, write $E(\mathbf{X})$ in terms of the new basis $\xi_1, \mathbf{d}_2, \ldots, \mathbf{d}_p$ for the estimation space. Now show that the least squares estimate of γ_1 equals the least squares estimate of β_1.]

9.8 Consider the linear model $\mathbf{X} \sim N(\beta_1\mathbf{d}_1 + \cdots + \beta_p\mathbf{d}_p, \sigma^2\mathbf{I})$ where the vectors $\mathbf{d}_1, \ldots, \mathbf{d}_p$ are linearly independent and $\mathbf{X}$ has n components. Consider testing for fixed i, $H_0\colon \beta_i = 0$ vs. $H_1\colon \beta_i \neq 0$, show that

$$\frac{\|\hat{\beta}_i\xi_i\|^2}{\mathrm{MS_e}} \sim F_{1,n-p}\left(\frac{\|\beta_i\xi_i\|}{\sigma}\right)$$

where ξ_i is that component of $\mathbf{d}_i$ which is orthogonal to the subspace spanned by the remaining columns of the design matrix. Conclude that any variable involved in "multicollinearity" tends to have a small F statistic since $\|\xi_i\|^2$ is small. The noncentrality parameter of the F statistic (assuming the correct model of course) is a function of $\|\xi_i\|$ so that for fixed β_i the power of the test is lowered by strong multicollinearity. Hint: without loss of generality let $i = 1$.

10 MULTIPLE REGRESSION ANALYSIS AND RELATED MATTERS

REGRESSION ANALYSIS

The methods we have used are also useful with other types of models. The factors we have been considering in our ANOVA models have taken on one of a fixed number of levels. Consider now a factor which might take values from a continuous range. Suppose that when the factor X is fixed, the expected value of the response Y varies linearly; that is, $E(Y|X) = \alpha + \beta X$. Suppose one observes X at several values, say, $x_1, \ldots, x_n$ and observes $y_1, \ldots, y_n$, the corresponding values of the response variable. This is shown schematically in Figure 10.1.

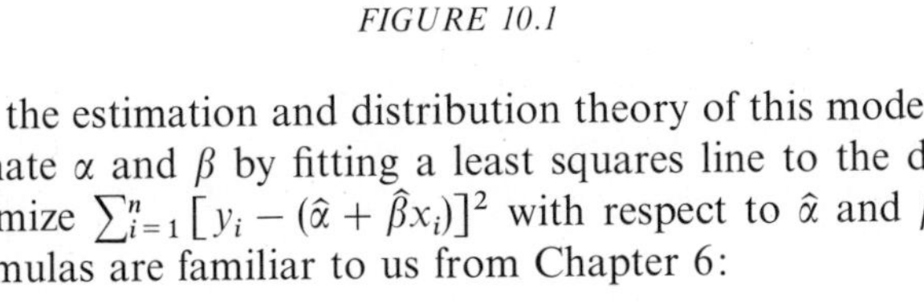

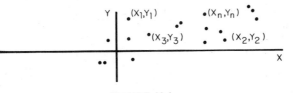

FIGURE 10.1

Consider the estimation and distribution theory of this model. We might try to estimate α and β by fitting a least squares line to the data, that is, try to minimize $\sum_{i=1}^{n} [y_i - (\hat{\alpha} + \hat{\beta}x_i)]^2$ with respect to $\hat{\alpha}$ and $\hat{\beta}$. The least squares formulas are familiar to us from Chapter 6:

$$E\left(\begin{pmatrix} Y_1 \\ \vdots \\ Y_n \end{pmatrix}\right) = \begin{pmatrix} 1 & x_1 \\ 1 & x_2 \\ 1 & x_3 \\ \vdots & \vdots \\ 1 & x_n \end{pmatrix} \begin{pmatrix} \alpha \\ \beta \end{pmatrix}$$

where the design matrix is

$$\mathbf{D} = \begin{pmatrix} 1 & x_1 \\ \vdots & \vdots \\ 1 & x_n \end{pmatrix};$$

$$\mathbf{D}'\mathbf{D} = \begin{pmatrix} 1 & \cdots & 1 \\ x_1 & \cdots & x_n \end{pmatrix} \begin{pmatrix} 1 & x_1 \\ \vdots & \vdots \\ 1 & x_n \end{pmatrix} = \begin{pmatrix} n & n\overline{x} \\ n\overline{x} & \sum x_i^2 \end{pmatrix}.$$

The least squares estimates are given by

$$(\mathbf{D}'\mathbf{D})^{-1}\mathbf{D}'\mathbf{y} = (\mathbf{D}'\mathbf{D})^{-1} \begin{pmatrix} \sum y_i \\ \sum y_i x_i \end{pmatrix} = \frac{1}{n\sum(x_i - \overline{x})^2} \begin{pmatrix} \sum x_i^2 \sum y_i - n\overline{x} \sum x_i y_i \\ n\sum x_i y_i - n^2 \overline{x}\overline{y} \end{pmatrix}.$$

Now

$$\begin{aligned} \sum x_i^2 \sum y_i - n\overline{x}\sum x_i y_i &= n\sum(x_i - \overline{x})^2\overline{y} + n^2\overline{x}^2\overline{y} - n\overline{x}\sum x_i y_i \\ &= n\sum(x_i - \overline{x})^2\overline{y} + \overline{x}n(n\overline{x}\overline{y} - \sum x_i y_i). \end{aligned}$$

Using the notation

$$\hat{\text{cov}}(X, Y) = \frac{\sum_i (x_i - \overline{x})(y_i - \overline{y})}{n - 1} = \frac{\sum_i x_i y_i - n\overline{x}\overline{y}}{n - 1}$$

and

$$\hat{\text{var}}(X) = \frac{\sum_i (x_i - \overline{x})^2}{n - 1},$$

we see that

$$\hat{\alpha} = \overline{y} - \overline{x}\frac{\hat{\text{cov}}(X, Y)}{\hat{\text{var}}(X)}, \qquad \hat{\beta} = \frac{\hat{\text{cov}}(X, Y)}{\hat{\text{var}}(X)},$$

or the least squares or (*simple linear*) *regression line* is given by

$$y = \overline{y} - \overline{x}\frac{\hat{\text{cov}}(X, Y)}{\hat{\text{var}}(X)} + \frac{\hat{\text{cov}}(X, Y)}{\hat{\text{var}}(X)}x \quad \text{or} \quad y - \overline{y} = \frac{\hat{\text{cov}}(X, Y)}{\hat{\text{var}}(X)}(x - \overline{x}).$$

Thus, the least squares line passes through the "mean" $(\overline{x}, \overline{y})$ of the data and has slope $\hat{\text{cov}}(X, Y)/\hat{\text{var}}(X)$. Note that when the (x_i, y_i) are from a bivariate sample, $\hat{\text{cov}}(X, Y)$ is the sample covariance and $\hat{\text{var}}(X)$ is the sample variance (see Problem 10.1).

Suppose that for each value of x, Y is normally distributed with the same variance, that is, $Y|x \sim N(\alpha + \beta x, \sigma^2)$. $Y|x$ denotes the random variable Y for a fixed value of x. Then if the observations are independent,

$$\begin{pmatrix} Y_1 \\ \vdots \\ Y_n \end{pmatrix} \sim N\left(\mathbf{D}\begin{pmatrix} \alpha \\ \beta \end{pmatrix}, \sigma^2\mathbf{I}\right)$$

TABLE 10.1

Source	SS	df	E(MS)
Mean or intercept	$n\bar{y}^2$	1	$\sigma^2 + n\alpha^2$
Slope	$(\sum x_i y_i)^2/\sum x_i^2$	1	$\sigma^2 + \beta^2(\sum_i x_i^2)$
Error	$\sum_i(y_i - (\hat{\alpha} + \hat{\beta}x_i))^2$	$n-2$	σ^2

and our previous theory holds. For example, $\hat{\alpha}$ and $\hat{\beta}$ are maximum likelihood estimates. The least squares estimates are normally distributed with covariance matrix

$$\sigma^2(\mathbf{D}'\mathbf{D})^{-1} = \sigma^2 \begin{pmatrix} \dfrac{\sum x_i^2}{n\sum(x_i - \bar{x})^2} & \dfrac{-\bar{x}}{\sum(x_i - \bar{x})^2} \\ \dfrac{-\bar{x}}{\sum(x_i - \bar{x})^2} & \dfrac{1}{\sum(x_i - \bar{x})^2} \end{pmatrix}.$$

Since we are considering the x_i's as fixed or at least doing the analysis conditionally upon the x_i's, we might choose $\sum x_i = 0$ (on redefining x_i by $x_i - \bar{x}$). In this case $\hat{\alpha} = \bar{y}$, $\hat{\beta} = \sum x_i y_i/\sum x_i^2$, and the columns of the design matrix are now orthogonal. Thus,

$$\|\mathbf{y}\|^2 = \sum_i \left(y_i - \bar{y} - x_i \frac{\sum_j x_j y_j}{\sum_j x_j^2} + \bar{y} + x_i \frac{\sum_j x_j y_j}{\sum_j x_j^2}\right)^2$$

$$= \sum_i \left(y_i - \bar{y} - x_i \frac{\sum_j x_j y_j}{\sum_j x_j^2}\right)^2 + n\bar{y}^2 + \frac{(\sum x_i y_i)^2}{\sum x_i^2}$$

which gives rise to an ANOVA table (Table 10.1).

Example 10.1 [R. A. Brown, R. Seed, and R. J. O'Connor, A comparison of relative growth in *Cerastoderma* (=*Cardium*) *edule*, *Modiolus modiolus*, and *Mytilus edulis* (*Mollusca: Bivalvia*), *Journal of Zoology, London* **179**, 297–315, (1976)][1] Brown *et al.* studied relative growth in bivalved molluscs. They examined their data for evidence of differential growth between the various size parameters measured by testing each pair of size variables y and x for their fit to the allometric equation $y = Ax^b$, where A and b are constants. When two variables having the same units of measurement are related by a power function of the allometric type, viz. $y = Ax^b$, values of the exponent b greater than unity indicate that the dimension y is increasing relatively faster than is x (positive allometry), and values of b below unity indicate the reverse (negative allometry); a value of unity for b is said to describe an isometric relationship in which the relative rates of growth for the two variables are identical, thus maintaining geometric similarity

[1] The Zoological Society of London.

TABLE 10.2

Allometric Relationships for the Parameters Measured in the Three Species, and a Test of Significance of Deviation from the Theoretical Slope

y	x	Regression[a]	Slope	SE of slope	Intercept	Theoretical slope	t-Test	Probability[b]
Shell weight	Tissue wet weight	1	1.028	.02494	.62492	1.0	1.128	NS
		2	.892	.02164	−.51513	1.0	−4.977	.001
Shell weight	Tissue dry weight	1	.930	.02680	3.13582	1.0	−.973	NS
		2	.954	.02747	−3.23064	1.0	−1.684	NS
Shell weight	Shell width	1	3.039	.03399	−7.66339	3.0	1.157	NS
		2	.323	.00361	2.52854	.3	−2.898	.005
Shell weight	Shell height	1	3.392	.03885	−9.50514	3.0	10.080	.001
		2	.289	.00331	2.80925	.3	−13.381	.001
Shell weight	Shell length	1	3.390	.04555	−9.71415	3.0	8.559	.001
		2	.287	.00386	2.87481	.3	−11.984	.001
Shell width	Shell height	1	1.107	.01173	−.57783	1.0	9.118	.001
		2	.888	.00941	.56628	1.0	−11.889	.001
Shell width	Shell length	1	1.108	.01320	−.65125	1.0	8.182	.001
		2	.883	.01052	.64343	1.0	−11.090	.001
Shell length	Shell height	1	.993	.00687	.08465	1.0	−.990	NS
		2	1.000	.00692	−.06175	1.0	−.694	NS

[a] 1 = log y regressed on log x; 2 = log x regressed on log y.
[b] NS = not significant.

with size increase. If the dimensions of x are different from those of y, different criteria for allometry and isometry apply, for example, if y is a volume (L^3) and x is a length (L), then $b = 3$ corresponds to isometry. The allometric equation can be rewritten in linear form as $\log y = \log A + b \log x$. The constants $\log A$ and $\log b$ were estimated by the use of least squares regression. Since in most of the bivariate plots neither variable had any special claim to treatment as the independent variable, both the regressions of $\log y$ on $\log x$ and of $\log x$ on $\log y$ were calculated. The sample of *C. edule* contained 184 animals. Estimated allometric relationships for the various parameters measured in *C. edule* along with a test of significance of deviation from the theoretical slope are presented in Table 10.2.

The F-test of the ANOVA table tests the hypothesis that $\beta = 0$. To test that $\beta = \beta_0 \neq 0$ a t-test is used. One knows that

$$\mathrm{var}(\hat{\beta}) = \sigma^2 / \sum (x_i - \bar{x})^2$$

by taking the appropriate element of $\sigma^2(\mathbf{D}'\mathbf{D})^{-1}$. Let s^2 be the residual mean square; then the t-test is

$$t = \frac{(\hat{\beta} - \beta_0)\sqrt{\sum (x_i - \bar{x})^2}}{s}.$$

The standard deviation of an estimate is called the standard error (SE).

Note that the regression of shell weight on tissue wet weight is not significant, while the regression of tissue wet weight on shell weight is significant ($p < .001$). Since there was no "independent variable," the authors averaged the estimated slope coefficients from the regressions of $\log y$ on $\log x$ and of $\log x$ on $\log y$ and then drew their conclusions as to whether growth was allometric. For example, the authors concluded that shell weight was isometrically related to tissue wet weight and to tissue dry weight in *C. edule*.

Example 10.2 [B. Wallace, L. Fisher, and J. Tremann, Lymphocyte culture studies in patients with prostatic carcinoma, unpublished manuscript] The cancer stage was graded as *A*, *B*, *C*, or *D* in 29 men with cancer of the prostate. *A* was the earliest stage and *D* the most developed stage. Stimulation of the patients' lymphocytes with phytohemagglutinin (PHA) causes blastogenesis formation. The relative amount of blastogenesis with and without PHA in sera from the cancer patient and a pooled normal control sera was investigated in its relationship to the stage of the cancer. The variable used in Table 10.3 is positive if the patient responded less well to PHA in his own sera than with the control sera. The data and tumor stage are coded as 1, 2, 3, and 4. To look for association, Y was regressed on X (see Figure 10.2). Note that (4, 7.27) is an outlying observation. In this case, even with the outlier the regression line (given as a solid line) does not have a slope that

TABLE 10.3

Tumor stage, Y	Comparative amount of stimulation, Y
A, $X = 1$	$-1.05, -.42, -.04, .68, .91$
B, $X = 2$	$-1.64, -.95, -.69, .27, .83, 1.16$
C, $X = 3$	$-.11, .02, .30, .61, .75, .89, 2.00, 2.17$
D, $X = 4$	$-2.60, -1.70, -.79, .12, .27, .35, .51, .77, 1.34, 7.27$

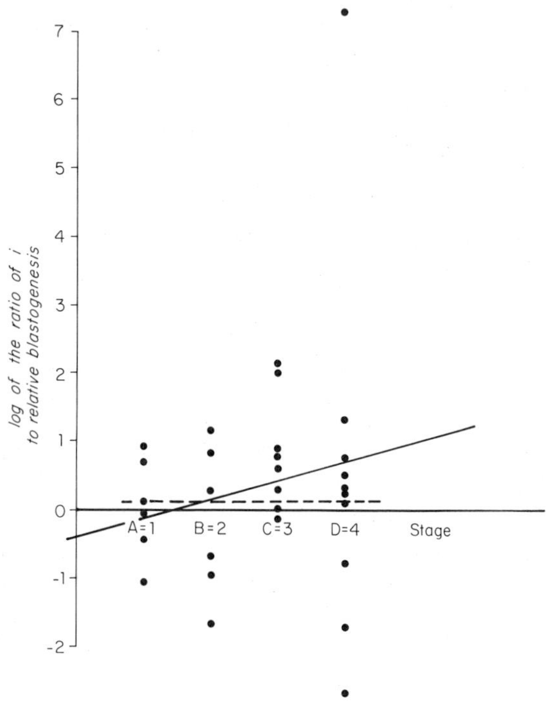

FIGURE 10.2 Slope of the regression line is not significant even with the large outlying observation (D, 7.27) included in the analysis. $r = .16$, $Y = -.31 + .25X$. Omitting the outlying observation, $r = .004$, $Y = .13 + .004X$.

is significantly nonzero. However, on removing the outlying observation, the slope of the regression line is almost identically zero, and the lack of a relationship is even clearer.

DEFINITION 10.1 Let X and Y be two jointly distributed random variables with finite positive variances; the *correlation* of X and Y is

$$\rho = \frac{\operatorname{cov}(X, Y)}{\sqrt{\operatorname{var}(X)}\sqrt{\operatorname{var}(Y)}}.$$

ρ is also called the *correlation coefficient.*

The following are facts about the correlation coefficient.

(1) $-1 \le \rho \le 1$ follows from the Cauchy–Schwarz inequality.
(2) If X and Y are independent, then $\rho = 0$.
(3) If $\rho = 0$ and X and Y are bivariate normal, then X and Y are independent.

LEMMA 10.1 Let $\operatorname{var}(X) > 0$ and $\operatorname{cov}(X, Y)$ exist. Then $\operatorname{var}(Y) = E((Y - \hat{Y})^2) + \beta^2 E((X - \mu_X)^2)$, where $\hat{Y} = \mu_Y + \beta(X - \mu_X)$ and $\beta = \operatorname{cov}(X, Y)/\operatorname{var}(X)$ where $\mu_X \equiv E(X)$ and $\mu_Y \equiv E(Y)$.

Proof

$$\begin{aligned} E((Y - \mu_Y)^2) &= E(((Y - \hat{Y}) + \beta(X - \mu_X))^2) \\ &= E((Y - \hat{Y})^2) + 2\underbrace{E((Y - \hat{Y})\beta(X - \mu_X))}_{=0} \\ &\quad + \beta^2 E((X - \mu_X)^2). \end{aligned} \qquad \square$$

Dividing by $\operatorname{var}(Y)$, we obtain

$$1 = \frac{E((Y - \hat{Y})^2)}{\operatorname{var}(Y)} + \rho^2 \qquad \text{or} \qquad \frac{E((Y - \hat{Y})^2)}{\operatorname{var}(Y)} = 1 - \rho^2.$$

$1 - \rho^2$ is the fraction of the Y variability that cannot be explained by linear prediction in X. Thus, ρ^2 is the fraction of the Y variability that can be explained by linear prediction in X.

For our least squares estimates from a sample, one gets the same sort of relationship. Let

$$r = \frac{\sum_{i=1}^n (X_i - \bar{X})(Y_i - \bar{Y})}{\sqrt{\sum_{i=1}^n (X_i - \bar{X})^2 \sum_{i=1}^n (Y_i - \bar{Y})^2}}$$

be the sample correlation, and let $\hat{Y}_i = \hat{\alpha} + \hat{\beta} X_i$ the least squares predicted value of Y_i. Then

$$1 = \frac{\sum_{i=1}^n (Y_i - \hat{Y}_i)^2}{\sum_{i=1}^n (Y_i - \bar{Y})^2} + r^2.$$

A *second interpretation* of ρ is obtained from standardizing the random variables X and Y to have mean zero and variance one:

$$\hat{Y} - \mu_Y = \beta(X - \mu_X) = \frac{\operatorname{cov}(X, Y)}{\operatorname{var}(X)}(X - \mu_X)$$

or

$$\frac{\hat{Y} - \mu_Y}{\sqrt{\operatorname{var}(Y)}} = \frac{\operatorname{cov}(X, Y)}{\sqrt{\operatorname{var}(X)}\sqrt{\operatorname{var}(Y)}} \frac{X - \mu_X}{\sqrt{\operatorname{var}(X)}}$$

or

$$\hat{Y}_{\text{standardized}} = \rho X_{\text{standardized}}.$$

The standardized variable is the linear transform that has mean zero and variance one. Thus, the correlation coefficient ρ is the slope of the least squares line between the standardized variables.

A *third interpretation of* ρ comes from the interpretation of $\operatorname{cov}(X, Y)$ as the inner product between X and Y. Thus, the inner product of X and Y, "$X'Y$," is $\|X\|\,\|Y\|\cos\theta_{X,Y}$ or

$$\cos\theta_{X,Y} = \frac{X'Y}{\|X\|\,\|Y\|} \qquad \text{or} \qquad \cos\theta_{X,Y} = \frac{\operatorname{cov}(X, Y)}{\sqrt{\operatorname{cov}(X, X)}\sqrt{\operatorname{cov}(Y, Y)}} = \rho.$$

The correlation coefficient may be thought of as the cosine of the angle between X and Y (when adjusted to have mean zero). In point of fact, r *is* the cosine of the angle between

$$\begin{pmatrix} x_1 - \bar{x} \\ x_2 - \bar{x} \\ \vdots \\ x_n - \bar{x} \end{pmatrix} \quad \text{and} \quad \begin{pmatrix} y_1 - \bar{y} \\ y_2 - \bar{y} \\ \vdots \\ y_n - \bar{y} \end{pmatrix}.$$

Before moving on to more regression topics, we discuss some theory associated with the multivariate normal distribution. As the multivariate normal distribution underlies much of multivariate statistical analysis, we will do this in some generality.

We consider only the case of positive-definite covariance matrices. Let

$$\mathbf{X} \sim N(\boldsymbol{\mu}, \mathbf{C}) \qquad \text{where} \quad \mathbf{X} = \begin{pmatrix} X_1 \\ \vdots \\ X_n \end{pmatrix}.$$

Partition $\mathbf{X}^{n \times 1}$ by

$$\mathbf{X}_1 = \begin{pmatrix} X_1 \\ \vdots \\ X_k \end{pmatrix} \qquad \text{and} \qquad \mathbf{X}_2 = \begin{pmatrix} X_{k+1} \\ \vdots \\ X_n \end{pmatrix}.$$

Corresponding to this partition of $\mathbf{X}$, we divide $\boldsymbol{\mu}$ and $\mathbf{C}$,

$$\boldsymbol{\mu} = \begin{pmatrix} \boldsymbol{\mu}_1 \\ \boldsymbol{\mu}_2 \end{pmatrix}, \qquad \mathbf{C}^{n \times n} = \underset{k \text{ cols} \quad n-k \text{ cols}}{\left(\begin{array}{c|c} \mathbf{C}_{11} & \mathbf{C}_{12} \\ \hline \mathbf{C}_{21} & \mathbf{C}_{22} \end{array}\right)} \begin{array}{l} k \text{ rows} \\ n-k \text{ rows} \end{array}.$$

THEOREM 10.1 If $\mathbf{X}$ is as defined above, then the marginal distribution of $\mathbf{X}_1$ is $N(\boldsymbol{\mu}_1, \mathbf{C}_{11})$.

Proof Problem 10.6. □

THEOREM 10.2 Let

$$\mathbf{X} \sim N(\boldsymbol{\mu}, \mathbf{C}), \qquad \mathbf{X} = \begin{pmatrix} \mathbf{X}_1 \\ \mathbf{X}_2 \end{pmatrix}.$$

The conditional distribution of $\mathbf{X}_1$, given that $\mathbf{X}_2 = \mathbf{x}_2$, is

$$N(\boldsymbol{\mu}_1 + \mathbf{C}_{12}\mathbf{C}_{22}^{-1}(\mathbf{x}_2 - \boldsymbol{\mu}_2), \mathbf{C}_{11} - \mathbf{C}_{12}\mathbf{C}_{22}^{-1}\mathbf{C}_{21}).$$

Proof Problem 10.7. □

As an application, let $(X_1, Y_1), (X_2, Y_2), \ldots, (X_n, Y_n)$ be a sample from the bivariate normal distribution

$$N\left(\begin{pmatrix} \mu_X \\ \mu_Y \end{pmatrix}, \begin{pmatrix} \sigma_X{}^2 & \sigma_{12} \\ \sigma_{12} & \sigma_Y{}^2 \end{pmatrix}\right).$$

Let us find the conditional distribution of

$$\mathbf{Y} = \begin{pmatrix} Y_1 \\ \vdots \\ Y_n \end{pmatrix}, \qquad \textit{given}\text{ that} \qquad \mathbf{X} = \begin{pmatrix} X_1 \\ \vdots \\ X_n \end{pmatrix}.$$

Let

$$\mathbf{Z} = \begin{pmatrix} Z_1 \\ \vdots \\ Z_{2n} \end{pmatrix} = \begin{pmatrix} Y_1 \\ \vdots \\ Y_n \\ X_1 \\ \vdots \\ X_n \end{pmatrix}.$$

Let $\mathbf{I}$ be the $n \times n$ identity matrix. Then it is easy to see that $\mathbf{Z}$ is $N(\boldsymbol{\mu}, \mathbf{C})$, where

$$\boldsymbol{\mu} = \begin{pmatrix} \mu_Y \\ \vdots \\ \mu_Y \\ \mu_X \\ \vdots \\ \mu_X \end{pmatrix} \begin{matrix} \left.\vphantom{\begin{matrix} a \\ b \\ c \end{matrix}}\right\} n \text{ times} \\ \left.\vphantom{\begin{matrix} a \\ b \\ c \end{matrix}}\right\} n \text{ times} \end{matrix}, \qquad \mathbf{C} = \left(\begin{array}{c|c} \sigma_Y{}^2\mathbf{I} & \sigma_{12}\mathbf{I} \\ \hline \sigma_{12}\mathbf{I} & \sigma_X{}^2\mathbf{I} \end{array}\right).$$

Using Theorem 10.2, we see that $\mathbf{Y}$, given $\mathbf{X} = \mathbf{x}$, is normal with mean

$$\boldsymbol{\mu}_1 + \mathbf{C}_{12}\mathbf{C}_{22}^{-1}(\mathbf{x} - \boldsymbol{\mu}_2) = \begin{pmatrix} \mu_Y \\ \vdots \\ \mu_Y \end{pmatrix} + \frac{\sigma_{12}}{\sigma_X{}^2} \begin{pmatrix} x_1 - \mu_X \\ \vdots \\ x_n - \mu_X \end{pmatrix}$$

and covariance matrix

$$\mathbf{C}_{11} - \mathbf{C}_{12}\mathbf{C}_{22}^{-1}\mathbf{C}_{21} = \sigma_Y{}^2\mathbf{I} - \sigma_{12}\frac{1}{\sigma_X{}^2}\sigma_{12}\mathbf{I}$$

$$= (\sigma_Y{}^2 - \sigma_{12}^2/\sigma_X{}^2)\mathbf{I} = \sigma_Y{}^2(1 - \rho^2)\mathbf{I}.$$

Thus, if we set

$$\alpha = \mu_Y - \frac{\sigma_{12}}{\sigma_X{}^2}\mu_X, \qquad \beta = \frac{\sigma_{12}}{\sigma_X{}^2}, \qquad \sigma^2 = \sigma_Y{}^2(1 - \rho^2),$$

then

$$Y|X \sim N(\begin{pmatrix} 1 & X_1 \\ \vdots & \vdots \\ 1 & X_n \end{pmatrix}\begin{pmatrix} \alpha \\ \beta \end{pmatrix}, \sigma^2\mathbf{I}),$$

or the setup appropriate for our regression model. Therefore, given bivariate normal data, linear regression is appropriate, the analysis being thought of as conditional upon the observed values of the X's. Note the nice partition of $\sigma_Y{}^2 = \sigma^2 + \rho^2\sigma_Y{}^2$, where $(1 - \rho^2)\sigma_Y{}^2$ is the variability left about the regression line.

MULTIPLE REGRESSION

The least squares analysis carried out previously allows us to use several predictor or independent X variables. One also can use more complex functions of X than linear functions to predict Y *provided* the unknown parameters enter linearly. We now present examples of situations in which more complicated regression models are used.

Polynomial Regression

When predicting values Y_i with a regression equation, let $\hat{Y}_i$ be the value on the *regression curve* predicted for Y_i. The values $Y_i - \hat{Y}_i$ are called the residuals. Suppose we carried out a simple linear regression and plotted the residuals $Y_i - \hat{Y}_i$ versus X_i and obtained a plot like that shown in Figure 10.3. We tend to have values greater than zero in the middle of the X range and less than zero at the endpoints [Note: $\sum_i (Y_i - \hat{Y}_i) = 0$ (Problem 10.8).] This indicates that a straight line does not fit the data well; thus, we might assume or try the model

$$E(Y|X) = \beta_0 + \beta_1 X + \beta_2 X^2.$$

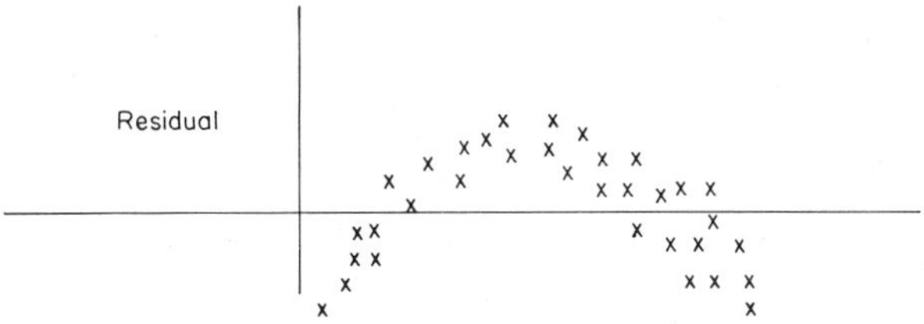

FIGURE 10.3

The design matrix **D** will look like

$$\mathbf{D} = \begin{pmatrix} 1 & X_1 & X_1{}^2 \\ 1 & X_2 & X_2{}^2 \\ \vdots & \vdots & \vdots \\ 1 & X_n & X_n{}^2 \end{pmatrix}.$$

Then, as usual, $\hat{\boldsymbol{\beta}} = (\mathbf{D}'\mathbf{D})^{-1}\mathbf{D}'\mathbf{Y}$ is the least squares estimate and maximum likelihood estimate and $\mathrm{SS_e} = \|\mathbf{Y} - \mathbf{D}\hat{\boldsymbol{\beta}}\|^2$. This model is a second-order *polynomial regression* model.

Multiple Regression

Suppose we have several variables affecting the outcome Y. Let $E(Y|X_1, \ldots, X_n) = \beta_0 + \beta_1 X_1 + \cdots + \beta_n X_n$. This is called a multiple regression equation, “multiple” in that multiple predictor variables are used.

Periodic Regression

Suppose the data Y_i are observed at times T_i. If there is cyclic (yearly) behavior (T_i in days), one might model the system by the first few terms in a Fourier series expansion, for example,

$$E(Y|t) = \beta_0 + \beta_1 \sin\left(\frac{t2\pi}{365}\right) + \beta_2 \cos\left(\frac{t2\pi}{365}\right) + \beta_3 \sin\left(\frac{t4\pi}{365}\right) + \beta_4 \cos\left(\frac{t4\pi}{365}\right).$$

In this case, the design matrix is

$$\mathbf{D} = \begin{pmatrix} 1 & \sin\dfrac{t_1 2\pi}{365} & \cos\dfrac{t_1 2\pi}{365} & \sin\dfrac{t_1 4\pi}{365} & \cos\dfrac{t_1 4\pi}{365} \\ \vdots & \vdots & \vdots & \vdots & \vdots \\ 1 & \sin\dfrac{t_n 2\pi}{365} & \cos\dfrac{t_n 2\pi}{365} & \sin\dfrac{t_n 4\pi}{365} & \cos\dfrac{t_n 4\pi}{365} \end{pmatrix}.$$

This is an example of a regression approach to a time series problem. Of course, we need independent normal errors at each t_i with the same variance.

Consider now the mathematics of these more complex regression situations. The least squares line passes through the mean of the distribution or data. This is shown by the following two assertions.

LEMMA 10.2 (a) For jointly distributed $Y, X_1, \ldots, X_k$,

$$\min_{\beta_0, \ldots, \beta_k} E([Y - (\beta_0 + \beta_1 X_1 + \cdots + \beta_k X_k)]^2)$$

$$= \min_{\beta_1, \ldots, \beta_k} E([(Y - \mu_Y) - (\beta_1(X_1 - \mu_{X_1}) + \cdots + \beta_k(X_k - \mu_{X_k}))]^2)$$

(that is, $\beta_0 = +\mu_Y - \beta_1\mu_{X_1} - \cdots - \beta_k\mu_{X_k}$).

(b) For the data $(Y_i, X_{i1}, \ldots, X_{ik})$, $i = 1, \ldots, n$,

$$\min_{\beta_0, \ldots, \beta_k} \sum_{i=1}^{n} (Y_i - (\beta_0 + \beta_1 X_{i1} + \cdots + \beta_k X_{ik}))^2$$

$$= \min_{\beta_1, \ldots, \beta_k} \sum_{i=1}^{n} ((Y_i - \bar{Y}) - (\beta_1(X_{i1} - \bar{X}_1) + \cdots + \beta_k(X_{ik} - \bar{X}_k)))^2$$

(where $\bar{X}_j = (1/n)\sum_{i=1}^{n} X_{ij}$, $\bar{Y} = (1/n)\sum_{i=1}^{n} Y_i$).

Proof Problem 10.9. □

THEOREM 10.3 Let $Y, X_1, \ldots, X_n$ be jointly distributed with mean vector and positive-definite covariance matrix

$$\begin{pmatrix} \mu_Y \\ \boldsymbol{\mu}_{\mathbf{X}} \end{pmatrix} \quad \text{and} \quad \mathbf{C} = \left(\begin{array}{c|c} C_{11} & \mathbf{C}_{12} \\ \hline \mathbf{C}_{21} & \mathbf{C}_{22} \end{array}\right),$$

respectively. The least squares linear predictor of Y in terms of $X_1, \ldots, X_n$ is given by $\hat{Y} = \mu_Y + \mathbf{C}_{12}\mathbf{C}_{22}^{-1}(\mathbf{X} - \boldsymbol{\mu}_{\mathbf{X}})$; that is, $\min E((Y - (\beta_0 + \beta_1 X_1 + \cdots + \beta_n X_n))^2)$ is achieved when $\beta_0 + \cdots + \beta_n X_n = \mu_Y + \mathbf{C}_{12}\mathbf{C}_{22}^{-1}(\mathbf{X} - \boldsymbol{\mu}_{\mathbf{X}})$.

Proof If we know that the regression line passes through the mean (Problem 10.9), the result would follow by taking the projection of the data onto the subspace spanned by $X_1, \ldots, X_n$. We shall give a direct proof, however. Let $\hat{Y}$ be as above and $\tilde{Y} = a_0 + \sum_{i=1}^{n} a_i X_i$. Then $\hat{Y} - \tilde{Y} = b_0 + \sum_i b_i(X_i - \mu_i)$ (for some b_i's).

$$\begin{aligned} E((Y - \tilde{Y})^2) &= E((Y - \hat{Y} + \hat{Y} - \tilde{Y})^2) \\ &= E((Y - \hat{Y})^2) + E((\hat{Y} - \tilde{Y})^2) + 2E((Y - \hat{Y})(\hat{Y} - \tilde{Y})). \end{aligned}$$

The proof is complete if we show that $E((Y - \hat{Y})(\hat{Y} - \tilde{Y})) = 0$. This expected value equals

$$\underset{\substack{\| \\ 0}}{b_0 E(Y - \hat{Y}))} + \sum_{i=1}^{n} b_i\{E((Y - \mu_Y)(X_i - \mu_i) - \mathbf{C}_{12}\mathbf{C}_{22}^{-1}(\mathbf{X} - \boldsymbol{\mu}_\mathbf{X})(X_i - \mu_i))\}$$

$$= \sum_{i=1}^{n} b_i\left\{(\mathbf{C}_{12})_{1i} - \mathbf{C}_{12}\mathbf{C}_{22}^{-1}\begin{pmatrix}(\mathbf{C}_{22})_{1i} \\ \vdots \\ (\mathbf{C}_{22})_{ni}\end{pmatrix}\right\} = \sum_{i=1}^{n} b_i\{(\mathbf{C}_{12})_{1i} - \mathbf{C}_{12}\boldsymbol{\delta}_i\} = 0,$$

since $\mathbf{C}_{22}^{-1}$ times the ith column of $\mathbf{C}_{22}$ gives a vector $\boldsymbol{\delta}_i$ which is all zeros except for a one in the ith position. □

What is the geometric reasoning behind this proof?

THEOREM (COROLLARY) 10.4 Let $(y_i, x_{i1}, \ldots, x_{ik})$, $i = 1, \ldots, n$, be n $(k + 1)$-dimensional data points. Then

$$\min_{\beta_0, \ldots, \beta_k} \sum_{i=1}^{n} (y_i - (\beta_0 + \cdots + \beta_k x_{ik}))^2$$

is achieved when

$$\beta_0 + \beta_1 x_{i1} + \cdots + \beta_k x_{ik} = \bar{y} + \hat{\mathbf{C}}_{12}\hat{\mathbf{C}}_{22}^{-1}\begin{pmatrix} x_{i1} - \bar{x}_1 \\ \vdots \quad\quad \vdots \\ x_{ik} - \bar{x}_k\end{pmatrix},$$

where

$$\bar{y} = \frac{1}{n}\sum_{i=1}^{n} y_i, \qquad \bar{x}_j = \frac{1}{n}\sum_{i=1}^{n} x_{ij}, \qquad \hat{\mathbf{C}} = \left(\begin{array}{c|c}\hat{\mathbf{C}}_{11} & \hat{\mathbf{C}}_{12} \\ \hat{\mathbf{C}}_{21} & \hat{\mathbf{C}}_{22}\end{array}\right),$$

where "sample covariances" are used to construct $\hat{\mathbf{C}}$.

Proof Consider a probability measure putting mass $1/n$ at each $(y_i, x_{i1}, \ldots, x_{ik})$. Then

$$E((Y - (\beta_0 + \cdots + \beta_k X_k))^2) = \frac{1}{n}\sum_{i=1}^{n} (y_i - (\beta_0 + \cdots + \beta_k x_{ik}))^2.$$

Use the preceding theorem! In this case, $\widehat{\text{cov}}(X_i, X_j)$ has a $1/n$ multiplier instead of $1/(n - 1)$, but $\hat{\mathbf{C}}_{12}\hat{\mathbf{C}}_{22}^{-1}$ "cancels out" the difference of the constant factors. □

DEFINITION 10.2 The *multiple correlation coefficient* between Y and $(X_1, \ldots, X_k)$ is the correlation coefficient of Y and the best linear predictor of Y in terms of the X's. This is often denoted by R.

LEMMA 10.3

$$R^2 = 1 - \frac{E((Y - \hat{Y})^2)}{\mathrm{var}(Y)};$$

that is, "R^2 is the fraction of the Y variability linearly explained by $X_1, \ldots, X_k$."

Proof From the proof of Theorem 10.3 (with $\tilde{Y} \equiv 0$), $E((Y - \hat{Y})\hat{Y}) = 0$, $E(Y - \hat{Y}) = 0$. Thus,

$$\mathrm{cov}(Y, \hat{Y}) = \mathrm{cov}(Y - \hat{Y} + \hat{Y}, \hat{Y}) = \mathrm{cov}(Y - \hat{Y}, \hat{Y}) + \mathrm{var}(\hat{Y}) = \mathrm{var}(\hat{Y}),$$

$$\mathrm{var}(Y) = E((Y - \mu_Y)^2) = E((Y - \hat{Y} + \hat{Y} - E(\hat{Y}))^2) = E((Y - \hat{Y})^2) + \mathrm{var}(\hat{Y}),$$

$$R^2 = \frac{\mathrm{cov}^2(Y, \hat{Y})}{\mathrm{var}(Y)\,\mathrm{var}(\hat{Y})} = \frac{\mathrm{var}(\hat{Y})}{\mathrm{var}(Y)} = \frac{\mathrm{var}(Y) - E((Y - \hat{Y})^2)}{\mathrm{var}(Y)}. \quad \square$$

If we have sample data, R^2 is estimated by

$$\hat{R}^2 = \frac{(n-1)^{-1} \sum_{i=1}^{n} (\hat{Y}_i - \bar{Y}_i)^2}{(n-1)^{-1} \sum_{i=1}^{n} (Y_i - \bar{Y})^2}.$$

Even when the X_{ij}'s are fixed by the experimenter, we will talk about the "multiple correlation." The value of R^2 then depends on the choice of the X_{ij}'s. In the literature, R^2 is usually used in place of $\hat{R}^2$.

TABLE 10.4 ANOVA About Mean, $x_{ij} \equiv X_{ij} - \bar{X}_j$

Source	SS	df	E(MS)
Grand mean or intercept	$n\bar{Y}^2$	1	
Multiple regression or R^2	$\sum_i (\hat{Y}_i - \bar{Y})^2$	k	$\sigma^2 + \sum_i (\beta_i x_{i1} + \cdots + \beta_k x_{ik})^2/k$
Error or residual variability	$\sum_i (Y_i - \hat{Y}_i)^2$	$n-k-1$	σ^2
Total	$\sum_i Y_i^2$	n	

We may test that R^2 is zero by using our ANOVA tables (Table 10.4) and Cochran's theorem (k variables, n observations):

$$\sum Y_i^2 = n\bar{Y}^2 + \sum_i (Y_i - \hat{Y}_i)^2 + \sum_i (\hat{Y}_i - \bar{Y})^2.$$

To test $\beta_1 = \cdots = \beta_k = 0$, we use

$$F = \frac{\sum_i (\hat{Y}_i - \bar{Y})^2}{\sum_i (Y_i - \hat{Y}_i)^2} \frac{(n-k-1)}{k}. \quad \square$$

Now

$$1 - \hat{R}^2 = \sum_i (Y_i - \hat{Y}_i)^2 / \sum_i (Y_i - \bar{Y})^2$$

and

$$\hat{R}^2 = \frac{\sum_i (Y_i - \bar{Y})^2 - \sum_i (Y_i - \hat{Y}_i)^2}{\sum_i (Y_i - \bar{Y})^2} = \frac{\sum_i (\hat{Y}_i - \bar{Y})^2}{\sum_i (Y_i - \bar{Y})^2}.$$

Thus, the F-test uses the statistic

$$F = \frac{\hat{R}^2}{1 - \hat{R}^2} \frac{n - k - 1}{k}.$$

For the multivariate normal distribution, another interpretation of R^2 holds. From Theorem 10.2, we know that

$$\operatorname{var}(Y | X_1, \ldots, X_n) = (C_{11} - \mathbf{C}_{12}\mathbf{C}_{22}^{-1}\mathbf{C}_{21}).$$

We now relate this to $1 - R^2$.

THEOREM 10.5 $1 - R^2 = (C_{11} - \mathbf{C}_{12}\mathbf{C}_{22}^{-1}\mathbf{C}_{21})/C_{11}$.

Proof

$$C_{11} = \operatorname{var}(Y), \qquad 1 - R^2 = \frac{E((Y - \hat{Y})^2)}{C_{11}},$$

$$E((Y - \hat{Y})^2) = E(((1, -\beta_1, \ldots, -\beta_k)\begin{pmatrix} Y - \mu_Y \\ X_1 - \mu_1 \\ \vdots \\ X_k - \mu_k \end{pmatrix})2)$$

$$= \operatorname{var}((1, -\boldsymbol{\beta}')\begin{pmatrix} Y - \mu_Y \\ \vdots \\ X_k - \mu_k \end{pmatrix}) = (1, -\boldsymbol{\beta}')\mathbf{C}\begin{pmatrix} 1 \\ -\boldsymbol{\beta} \end{pmatrix}$$

$$= C_{11} - 2\boldsymbol{\beta}'\mathbf{C}_{21} + \boldsymbol{\beta}'\mathbf{C}_{22}\boldsymbol{\beta}.$$

From Theorem 10.3, we see that $\boldsymbol{\beta}' = \mathbf{C}_{12}\mathbf{C}_{22}^{-1}$. So

$$E((Y - \hat{Y})^2) = C_{11} - 2\mathbf{C}_{12}\mathbf{C}_{22}^{-1}\mathbf{C}_{21} + \mathbf{C}_{12}\mathbf{C}_{22}^{-1}\mathbf{C}_{22}\mathbf{C}_{22}^{-1}\mathbf{C}_{21}$$
$$= C_{11} - \mathbf{C}_{12}\mathbf{C}_{22}^{-1}\mathbf{C}_{21}. \quad \square$$

Example 10.3 [M. J. Gardner, Using the environment to explain and predict mortality, *Journal of the Royal Statistical Society A* **136**, Part 3, 421 (1973)] It is well known that death rates vary from area to area, year to year, and among different groups of the population. Gardner studied the relationship between a number of socioenvironmental factors and mortality. The study was limited to the County Boroughs of England and Wales. Death rates were calculated for all sizeable causes of death in middle age.

Four social and environmental factors (social factor score, "domestic" air pollution, latitude, and water calcium) were used as regressor variables in a multiple regression analysis of each death rate. The ordinary linear additive model was used in an attempt to explain absolute differences in the death rates. Other models were tried by transforming the dependent and/or the regressor variables, but these models produced broadly similar results. The relationship of the residuals to other environmental indices available was studied. The only factors showing sizeable and consistent correlation coefficients with the residuals were long-period average rainfall and longitude, those with rainfall being higher, for all causes of death. Rainfall was therefore included as a fifth regressor variable. A repeat examination of the "new" residuals showed no other factor, for which data are available, to be important.

The fitted regression coefficients for the standardized variables (see table footnote) are given in Table 10.5. Note the high values of $\hat{R}^2$ (% of variance explained).

TABLE 10.5

Multiple Regression[a] of Local Death Rates on Five Socioenvironmental Indices in the County Boroughs

Sex–age group	Period	Social factor score	"Domestic" air pollution	Latitude	Water calcium	Long period average rainfall	% of variance explained
Males 45–64	1948–1954	+.16	+.48***	+.10	−.23*	+.27***	80
	1958–1964	+.19*	+.36***	+.21**	−.24**	+.30***	84
Males 65–74	1950–1954	+.24*	+.28*	+.02	−.43***	+.17	73
	1958–1964	+.39**	+.17	+.13	−.30**	+.21	76
Females 45–64	1948–1954	+.16	+.20	+.32**	−.15	+.40***	73
	1958–1964	+.29*	+.12	+.19	−.22*	+.39***	72
Females 65–74	1950–1954	+.39***	+.02	+.36***	−.12	+.40***	80
	1958–1964	+.40**	−.05	+.29***	−.27**	+.29**	73

* $p < .05$. ** $p < .01$. *** $p < .001$.

[a] Standardized partial regression coefficients given, that is, the variables are reduced to the same mean (zero) and variance (one) to allow values for the five socioenvironmental indices in each cause of death to be compared. The higher of two coefficients is not necessarily the more significant statistically.

The regression equations were developed with data from 61 English County Boroughs. The equations were then used to predict values for 12 other County Boroughs. The square of the correlation coefficients for the predicted and observed values are presented in Table 10.6.

TABLE 10.6

Results of Using Estimated Multiple Regression Equations from 61 County Boroughs to Predict Death Rates in 12 Other County Boroughs

Sex–age group	Period	$\hat{R}^2$	r^2 [a]
Males 45–64	1948–1954	.80	.12
	1958–1964	.84	.26
Males 65–74	1950–1954	.73	.09
	1958–1964	.76	.25
Females 45–64	1948–1954	.73	.46
	1958–1964	.72	.48
Females 65–74	1950–1954	.80	.53
	1958–1964	.73	.41

[a] r is the correlation coefficient in the second sample between the predicted value of the dependent variable and its observed value.

Note the dramatic drop in predictive ability when new Boroughs were used. In model building it is not unusual to have a large drop in predictability between a fitted model and new independent situations. The drop results from three factors:

(1) $E(\hat{R}^2) \geq R^2$.

(2) If one tries to fit many models, the multiple comparison problem enters in. One may get a very good fit by chance.

(3) The model is fitted in one environment and then used in another setting.

Many factors may be omitted that are relevant to prediction in the new setting but not in the old.

As a rule of thumb, the more complex the model, the less transportable the model is, in time and/or in space.

Example 10.4 [S. W. Coleman and K. Barth, Nutrient digestibility and N-metabolism by cattle fed rations based on urea and corn silage, *Journal of Animal Science* **39**, No. 2 (1974)] Multiple regression techniques were used by Coleman and Barth to determine the extent to which ration composition and intake factors are related to nutrient absorption and N- (nitrogen) utilization by steers fed natural rations composed of varying proportions of corn silage, urea, and shelled corn, and to determine the magnitude of these effects. Data were compiled from 55 steers in six different digestion and nitrogen-metabolism trials which were similar in manner of conduct.

Various coefficients of simple correlation involving digestion coefficients, measures of N-utilization, nutrient intake, and several other variables were calculated. The magnitude of these correlation coefficients and biological

TABLE 10.7

Statistical Analysis of Apparent Digestion Coefficients

Item	Apparent digestion coefficients		
	Dry matter (Y_a)	Nitrogen (Y_b)	Crude fiber (Y_c)
Coefficient of determination, R^2	.417	.311	.650
Mean square due to regression	76.39[b]	175.13[c]	789.92[b]
Mean square due to deviations from regression	8.54	31.03	33.97
Standard partial regression coefficients			
β_1	.017	.072	.117
β_2	.493	−.515	6.599
β_3	−.408	−.050	−.448
β_4	.082	−.175	.329
Standard error of partial regression coefficients for (df = 50) and			
β_1	.053, [t] = .10	.101, [t] = .41	.105, [t] = .94
β_2	.215, [t] = 3.03[c]	.410, [t] = 2.92[c]	.429, [t] = 4.76[b]
β_3	.037, [t] = 3.29[c]	.071, [t] = .37	.074, [t] = 4.66[b]
β_4	2.937, [t] = .65	5.597, [t] = 1.27	5.856, [t] = 3.36[c]
t values for testing β_i's = 0			
Regression equations			
$\hat{Y}_a = 75.73 + .005X_1 + .653X_2 - .123X_3 + 1.910X_4$			
$\hat{Y}_b = 82.60 + .041X_1 - 1.197X_2 - .027X_3 - 7.123X_4$			
$\hat{Y}_c = 60.64 + .099X_1 + 2.041X_2 - .347X_3 - 19.67X_4$			

[a] X_1, Percent N supplied by urea; X_2, energy (gross) supplied by concentrates (mcal/day); X_3, nitrogen intake (g/day); X_4, fiber intake (kg/day).

[b] $p < .001$. [c] $p < .01$.

importance were used to select variables for use in the multiple regression equations. Independent variables selected were

(1) X_1, percent N supplied by urea,
(2) X_2, energy supplied by concentrates (mcal/day),
(3) X_3, N intake (g/day), and
(4) X_4, fiber intake (kg/day).

A summary of statistical results for digestibility is presented in Table 10.7. The table presents the results for three different dependent variables.

In each case, the multiple regression was statistically significant with the individual terms (i.e., β_1, β_2, β_3, β_4) evaluated for statistical significance. To test $\beta_i = 0$ (in the multiple regression model), the authors used

$$t_i = \frac{\hat{\beta}_i}{\sqrt{\mathrm{MS_e}(\mathbf{D}'\mathbf{D})_{ii}^{-1}}}.$$

Why does this have a t-distribution under the null hypothesis that $\beta_i = 0$? How many degrees of freedom does the t variable have?

PARTIAL CORRELATION

Often in looking at a correlation coefficient, the interpretation is doubtful. Just because two variables are related, we are not justified in supposing that one "causes" the other. Often a person will argue that two variables are correlated because each is related to a third variable, and except for the relationship to the third variable they are not correlated. The partial correlation coefficient is an attempt to allow for the effect of intervening third variables.

DEFINITION 10.3 Let X and Y and $Z_1, \ldots, Z_k$ be jointly distributed random variables with finite covariance matrix. Let $\hat{X}$ be the best linear predictor for X in terms of $Z_1, \ldots, Z_k$ and $\hat{Y}$ the best linear predictor of Y in terms of $Z_1, \ldots, Z_k$. The *partial correlation coefficient* of X and Y adjusting for $Z_1, \ldots, Z_k$ is the correlation coefficient of $X - \hat{X}$ and $Y - \hat{Y}$.

To calculate the partial correlation $\rho_{X,Y \cdot Z_1, \ldots, Z_k}$, we take the joint covariance matrix of $X, Y, Z_1, \ldots, Z_k$:

$$\mathbf{C} = \begin{matrix} X \\ Y \\ Z_1 \\ \vdots \\ Z_k \end{matrix} \left(\begin{array}{c|c} \mathbf{C}_{11} & \mathbf{C}_{12} \\ \hline \mathbf{C}_{21} & \mathbf{C}_{22} \end{array} \right), \quad \text{columns } X\,Y\,\underbrace{Z_1, \ldots, Z_k}.$$

The 2×2 covariance matrix of $X - \hat{X}$ and $Y - \hat{Y}$ is denoted by

$$\begin{pmatrix} \sigma^2_{X \cdot Z_1,\ldots,Z_k} & \sigma_{X,Y \cdot Z_1,\ldots,Z_k} \\ \sigma_{X,Y \cdot Z_1,\ldots,Z_k} & \sigma^2_{Y \cdot Z_1,\ldots,Z_k} \end{pmatrix} = \mathbf{C}_{11} - \mathbf{C}_{12}\mathbf{C}_{22}^{-1}\mathbf{C}_{21}.$$

Then

$$\rho_{X,Y \cdot Z_1,\ldots,Z_k} = \frac{\sigma_{X,Y \cdot Z_1,\ldots,Z_k}}{\sqrt{\sigma^2_{X \cdot Z_1,\ldots,Z_k}\sigma^2_{Y \cdot Z_1,\ldots,Z_k}}}.$$

Since $\mathbf{C}_{11} - \mathbf{C}_{12}\mathbf{C}_{22}^{-1}\mathbf{C}_{21}$ is the correlation matrix of X and Y for *fixed* values $Z_1 = z_1, \ldots, Z_k = z_k$ when the $k + 2$ variables have a multivariate normal distribution, this gives another interpretation of the partial correlation coefficient.

$\rho_{X,Y \cdot Z_1,\ldots,Z_k}$ is the correlation *coefficient* of X and Y conditionally on knowing $Z_1, \ldots, Z_k$ (when we are dealing with multivariate normality).

Several other extensions along the line of the partial correlation coefficient are possible. In closing, we mention only two extensions:

(a) *Partial regression equations* The best linear predictor for Y in terms of X (when $Z_1, \ldots, Z_k$ are known).

(b) The *partial multiple correlation coefficient* This is the multiple correlation coefficient between Y and $X_1, \ldots, X_l$ when adjusting for $Z_1, \ldots, Z_k$ (denoted by $\rho_{Y(X_1,\ldots,X_l) \cdot Z_1,\ldots,Z_k}$). Here we subtract best linear predictors for Y and $X_1, \ldots, X_l$ in terms of $Z_1, \ldots, Z_k$ and then compute the multiple correlation coefficient of $Y - \hat{Y}$ and $X_1 - \hat{X}_1, \ldots, X_l - \hat{X}_l$.

PROBLEMS

10.1 Let $(X_1, Y_1), \ldots, (X_n, Y_n)$ be a sample from a bivariate distribution with finite variances and covariance. Show that

$$E\left(\sum_{i=1}^{n} (X_i - \bar{X})(Y_i - \bar{Y})/(n-1)\right) = \operatorname{cov}(X, Y)$$

and

$$E\left(\sum_{i=1}^{n} (X_i - \bar{X})^2/(n-1)\right) = \operatorname{var}(X).$$

10.2 Let $(X_1, Y_1), \ldots, (X_n, Y_n)$ be a bivariate normal sample:

$$\begin{pmatrix} X \\ Y \end{pmatrix} \sim N\left(\begin{pmatrix} \mu_1 \\ \mu_2 \end{pmatrix}, \begin{pmatrix} {\sigma_X}^2 & \rho\sigma_X\sigma_Y \\ \rho\sigma_X\sigma_Y & {\sigma_Y}^2 \end{pmatrix}\right).$$

Show that

(a) Conditionally upon $X_1 = x_1, \ldots, X_n = x_n$, the model of this chapter, $E(Y|X) = \alpha + \beta X$, holds.

(b) $\beta = 0 \leftrightarrow \rho = 0 \leftrightarrow \operatorname{cov}(X, Y) = 0; \leftrightarrow X$ and Y are independent normal random variables.

10.3 Let X and Y be jointly distributed random variables with finite positive variances. Show that $\min_{\alpha,\beta} E((Y - (\alpha + \beta X))^2)$ occurs when

$$\beta = \frac{\operatorname{cov}(X, Y)}{\operatorname{var}(X)} \qquad \text{and} \qquad \alpha = \mu_Y - \beta\mu_X.$$

10.4 The answer to Problem 10.3 has the same geometric reasoning as in our other cases; βX is the projection onto the X direction, that is:

Let $\mathbf{X} \sim N(\boldsymbol{\mu}^{n \times 1}, \mathbf{C})$, where $\mathbf{C}$ is positive definite. Each linear combination $\mathbf{a}'\mathbf{X}$ of entries of $\mathbf{X}$ defines a vector $\mathbf{V_a}$. Let $(\mathbf{V_a}, \mathbf{V_b}) \equiv \operatorname{cov}(\mathbf{a}'\mathbf{X}, \mathbf{b}'\mathbf{X})$.

(a) Show that $\{\mathbf{V_a}, (\cdot,\cdot) : \mathbf{a} \in R^n\}$ is an inner product space.

(b) If $\mathbf{X} = \binom{x}{y}$, then $\beta(x, 0) = (\operatorname{cov}(X, Y)/\operatorname{var}(X))(x, 0)$ is the projection of Y; that is, $(0, 1)\mathbf{X}$, onto the one-dimensional subspace generated by X, that is, $(1, 0)\mathbf{X}$.

(c) Thus, $\|Y - \beta X\|^2$ is minimized where $\beta = \operatorname{cov}(X, Y)/\operatorname{var}(X)$. But (show) $\|Y - \beta X\|^2 = E((Y - \mu_Y - \beta(X - \mu_X))^2)$.

10.5 Compute the slope and intercept of Example 10.2 with and without the "outlying" observation.

10.6 Prove Theorem 10.1. [*Hint*: How are marginal distributions related to joint moment generating functions?]

10.7 Prove Theorem 10.2. [*Hint*: (a) One only need show

$$f(\mathbf{x}_1, \mathbf{x}_2) = K(\mathbf{x}_2)\exp[-\tfrac{1}{2}(\mathbf{x}_1 - \text{mean})'(\text{cov matrix})^{-1}(\mathbf{x}_1 - \text{mean})].$$

(b) Show: If $\mathbf{A}$ is positive-definite symmetric and $\mathbf{B}$ is its inverse, where

$$\mathbf{A} = \begin{pmatrix} \mathbf{A}_{11} & \mathbf{A}_{12} \\ \mathbf{A}_{21} & \mathbf{A}_{22} \end{pmatrix}, \quad \mathbf{B} = \begin{pmatrix} \mathbf{B}_{11} & \mathbf{B}_{12} \\ \mathbf{B}_{21} & \mathbf{B}_{22} \end{pmatrix} \quad \begin{pmatrix} \text{same division of dimensions} \\ \text{with } \mathbf{A}_{11} \text{ square} \end{pmatrix}$$

then $\mathbf{A}_{11}^{-1} = \mathbf{B}_{11} - \mathbf{B}_{12}\mathbf{B}_{22}^{-1}\mathbf{B}_{21}$.

(c) $f(\mathbf{x}_1|\mathbf{x}_2) = f(\mathbf{x}_1, \mathbf{x}_2)/f_{\text{marginal}}(\mathbf{x}_2)$.]

10.8 Show that $\sum(y_i - \hat{y}_i) = 0$, where $\hat{y}_i$ is the predicted value of Y when $X = x_i$. $y_i - \hat{y}_i$ is called the residual. Thus, the residuals sum to zero in this context.

10.9 Prove Lemma 10.2. [*Hint*: Consider the correct answer plus any other constant.]

10.10 In the multiple regression problem, show that the following are equivalent:

(a) $\beta_1 = \beta_2 = \cdots = \beta_k = 0$;

(b) $R^2 = 0$.

10.11 Show that the equation following definition 10.3 does indeed give the covariance matrix of $X - \hat{X}$ and $Y - \hat{Y}$.

10.12 Prove that when n bivariate normal observations are taken and the least squares estimate $\hat{\beta}$ (of the slope of the regression line $E(Y|X) = \alpha + \beta X$ and sample correlation coefficient r are calculated, then a $(1 - \alpha)100\%$ confidence interval for β is given by

$$\left(\hat{\beta} - \hat{\beta}t_{\alpha/2,n-2}\frac{\sqrt{1-r^2}}{r\sqrt{n-2}}, \hat{\beta} + \hat{\beta}t_{\alpha/2,n-2}\frac{\sqrt{1-r^2}}{r\sqrt{n-2}}\right).$$

10.13 For bivariate observations X and Y, two models are considered (X is set or fixed):

(a) $Y_i = \beta_0 + \beta_1 X_i + e_i$, $i = 1,2,3,4$,
(b) $Y_i = \beta_0 + \beta_1 X_i + \beta_2 X_i^2 + e_i$, $i = 1,2,3,4$,

where the e_i are i.i.d. $N(0, \sigma^2)$.

From the following data, how would you test that model (a) holds?

i:	1	2	3	4
X_i:	−1	−2	1	2
Y_i:	3	2	1	2

10.14 Three variables X, Y, and Z have correlation coefficients $\rho_{X,Y}$, etc. Show that, when adjusting for Z, the partial correlation of X and Y satisfies

$$\rho_{X,Y\cdot Z} = \frac{\rho_{X,Y} - \rho_{X,Z}\rho_{Y,Z}}{\sqrt{(1-\rho_{X,Z}^2)(1-\rho_{Y,Z}^2)}}.$$

10.15 If $E(Y|X_1, \ldots, X_p) = \beta_0$ but we try fitting the model

$$E(Y|X_1, \ldots, X_p) = \beta_0 + \beta_1 X_1 + \cdots + \beta_p X_p,$$

show that under the true model $E(\hat{R}^2) = p(n-1)$, where n is the sample size. [*Hint*:

$$\hat{R}^2 = \frac{\|\hat{\mathbf{Y}} - \bar{Y}\mathbf{1}\|^2}{\|\mathbf{Y} - \bar{Y}\mathbf{1}\|^2} = \frac{\|(\mathbf{P}_M - \mathbf{P}_1)\mathbf{Y}\|^2}{\|(\mathbf{I} - \mathbf{P}_1)\mathbf{Y}\|^2} = \frac{\|(\mathbf{P}_M - \mathbf{P}_1)\mathbf{Y}\|^2}{\|(\mathbf{I} - \mathbf{P}_M)\mathbf{Y}\|^2 + \|(\mathbf{P}_M - \mathbf{P}_1)\mathbf{Y}\|^2},$$

where as usual $\mathbf{P}_M$ is the projection onto the estimation space and $\mathbf{P}_1$ is the projection onto the subspace spanned by the $\mathbf{1}$ vector. Note also that if $U \sim \chi_m^2$ and $V \sim \chi_n^2$ are independent, then $U/(U + V) \sim \text{beta}(m/2, n/2)$.]

APPENDIX 1 | REVIEW OF LINEAR ALGEBRA AND VECTOR SPACE THEORY

The following informal review of linear algebra and vector space theory will try to stress a geometric understanding of the material. The definition of certain concepts will not be given in full generality since our only interest is in making available to the student the minimum amount of background material needed to master this text. Basically, this text will require a knowledge of n-dimensional Euclidean space with the usual inner product.

A (real) *vector space* V is a nonempty set of objects, called vectors, together with two operations, one of which assigns to each pair of vectors $\mathbf{x}$ and $\mathbf{y}$ a vector $\mathbf{x} + \mathbf{y}$ called the sum of $\mathbf{x}$ and $\mathbf{y}$, and the other of which assigns to each real number α, called a scalar, and to each vector $\mathbf{x}$, a vector $\alpha\mathbf{x}$ called the product of α and $\mathbf{x}$. The first operation is called vector addition and is assumed to satisfy the following axioms for $\alpha, \beta \in \mathrm{R}$ (the reals) and for $\mathbf{x}, \mathbf{y}, \mathbf{z} \in V$:

(i) $\mathbf{x} + (\mathbf{y} + \mathbf{z}) = (\mathbf{x} + \mathbf{y}) + \mathbf{z}$ and $\mathbf{x} + \mathbf{y} = \mathbf{y} + \mathbf{x}$;

(ii) there exists a unique vector $\mathbf{0}$, called the zero vector, such that $\mathbf{x} + \mathbf{0} = \mathbf{x}$ for every vector $\mathbf{x}$;

(iii) for each vector $\mathbf{x}$ there is a unique vector $-\mathbf{x}$, called the negative of $\mathbf{x}$, such that $\mathbf{x} + (-\mathbf{x}) = \mathbf{0}$.

The second operation is called scalar multiplication and is assumed to satisfy the following axioms:

(i) $\alpha(\mathbf{x} + \mathbf{y}) = \alpha\mathbf{x} + \alpha\mathbf{y}$;

(ii) $(\alpha + \beta)\mathbf{x} = \alpha\mathbf{x} + \beta\mathbf{x}$;

(iii) $(\alpha\beta)\mathbf{x} = \alpha(\beta\mathbf{x})$;

(iv) $1\mathbf{x} = \mathbf{x}$.

DEFINITION 1 A vector $\mathbf{x}$ is a *linear combination* of vectors $\mathbf{v}_1, \ldots, \mathbf{v}_n$ if $\mathbf{x}$ can be expressed in the form $\mathbf{x} = \alpha_1\mathbf{v}_1 + \cdots + \alpha_n\mathbf{v}_n$ for some scalars $\alpha_1, \ldots, \alpha_n$.

DEFINITION 2 The finite set of vectors $\{\mathbf{v}_1, \ldots, \mathbf{v}_n\}$ is *linearly dependent* if there exists scalars $\alpha_1, \ldots, \alpha_n$, not all zero, such that $\alpha_1\mathbf{v}_1 + \cdots + \alpha_n\mathbf{v}_n = \mathbf{0}$. A finite set of vectors which is not linearly dependent is said to be *linearly independent*. Thus, $\{\mathbf{v}_1, \ldots, \mathbf{v}_n\}$ is a linearly independent set of vectors if and only if $\alpha_1\mathbf{v}_1 + \cdots + \alpha_n\mathbf{v}_n = \mathbf{0}$ implies $\alpha_1 = \cdots = \alpha_n = 0$.

THEOREM 1 A set of vectors is linearly dependent if and only if it is always possible to express at least one of them as a linear combination of the others.

THEOREM 2 Any subset of a linearly independent set is linearly independent.

THEOREM 3 Any set which contains a linearly dependent set is linearly dependent.

DEFINITION 3 A *basis or coordinate system* is a set of linearly independent vectors $\mathbf{v}_1, \ldots, \mathbf{v}_n$ such that every vector can be written as a linear combination of $\mathbf{v}_1, \ldots, \mathbf{v}_n$.

In general, there are many different possible bases for a given vector space so that a basis for a vector space is not unique. However, the number of vectors in any basis for a given vector space is unique and is called the *dimension* of the vector space. A vector space is finite dimensional if it has a finite basis. All the vector spaces we shall deal with are finite dimensional.

THEOREM 4 and DEFINITION 4 Any vector can be expressed as a unique linear combination of a given basis. Therefore, for a vector space with basis $\mathbf{v}_1, \ldots, \mathbf{v}_n$, any vector $\mathbf{x}$ can be written in the form $\mathbf{x} = \alpha_1\mathbf{v}_1 + \cdots + \alpha_n\mathbf{v}_n$ where the scalars α_i are uniquely determined by $\mathbf{x}$ and the basis $\mathbf{v}_1, \ldots, \mathbf{v}_n$ and are called the *coordinates* of $\mathbf{x}$ relative to this basis.

Example The set R^n of all ordered n-tuples

$$\mathbf{x} = \begin{pmatrix} x_1 \\ x_2 \\ \vdots \\ x_n \end{pmatrix}$$

of real numbers together with the usual operations of addition and multiplication of n-tuples by real numbers is a vector space. The n-tuples are called vectors. Recall that the vector space operations are defined componentwise by

$$\begin{pmatrix} x_1 \\ \vdots \\ x_n \end{pmatrix} + \begin{pmatrix} y_1 \\ \vdots \\ y_n \end{pmatrix} = \begin{pmatrix} x_1 + y_1 \\ \vdots \\ x_n + y_n \end{pmatrix} \quad \text{and} \quad \alpha \begin{pmatrix} x_1 \\ \vdots \\ x_n \end{pmatrix} = \begin{pmatrix} \alpha x_1 \\ \vdots \\ \alpha x_n \end{pmatrix}$$

and the zero vector is

$$\mathbf{0} = \begin{pmatrix} 0 \\ \vdots \\ 0 \end{pmatrix}.$$

Let

$$\boldsymbol{\delta}_1 = \begin{pmatrix} 1 \\ 0 \\ 0 \\ \vdots \\ 0 \end{pmatrix}, \quad \boldsymbol{\delta}_2 = \begin{pmatrix} 0 \\ 1 \\ 0 \\ \vdots \\ 0 \end{pmatrix}, \ldots, \boldsymbol{\delta}_n = \begin{pmatrix} 0 \\ 0 \\ 0 \\ \vdots \\ 1 \end{pmatrix}$$

where $\boldsymbol{\delta}_i$ has a 1 in the ith position and 0's elsewhere. Letting $x_1, x_2, \ldots, x_n$ be scalars, we have $\mathbf{x} = \sum_{i=1}^{n} x_i \boldsymbol{\delta}_i$ and it is easily seen that $\boldsymbol{\delta}_1, \ldots, \boldsymbol{\delta}_n$ are linearly independent so that the $\boldsymbol{\delta}_i$ form a basis or coordinate system. This particular basis is called the standard or canonical basis of R^n. It is important to note that the n coordinates of $\mathbf{x}$ with respect to the canonical basis are $x_1, x_2, \ldots, x_n$. Since there is a natural ordering of the vectors in the canonical basis, we call x_i the ith coordinate of $\mathbf{x}$ and note that it is found in the ith position of the n-tuple.

Example Let $\mathbf{b}_1 = \binom{1}{1}$ and $\mathbf{b}_2 = \binom{1}{-1}$ be another basis for R^2. We can express the vector $\binom{2}{1}$ as

$$\begin{pmatrix} 2 \\ 1 \end{pmatrix} = 2\begin{pmatrix} 1 \\ 0 \end{pmatrix} + 1\begin{pmatrix} 0 \\ 1 \end{pmatrix} \quad \text{or} \quad \begin{pmatrix} 2 \\ 1 \end{pmatrix} = \frac{3}{2}\begin{pmatrix} 1 \\ 1 \end{pmatrix} + \frac{1}{2}\begin{pmatrix} 1 \\ -1 \end{pmatrix},$$

that is,

$$\begin{pmatrix} 2 \\ 1 \end{pmatrix} = 2\boldsymbol{\delta}_1 + 1\boldsymbol{\delta}_2 \quad \text{or} \quad \begin{pmatrix} 2 \\ 1 \end{pmatrix} = \frac{3}{2}\mathbf{b}_1 + \frac{1}{2}\mathbf{b}_2.$$

Now the basis vectors $\mathbf{b}_1$ and $\mathbf{b}_2$ have no natural ordering; we will thus just use their arbitrary subscript ordering so that we can talk about the ith coordinate of $\binom{2}{1}$ relative to the basis $\mathbf{b}_1$ and $\mathbf{b}_2$. The first coordinate of $\binom{2}{1}$ relative to the canonical basis is 2 and is found in the first position of the 2-tuple; the second coordinate of $\binom{2}{1}$ relative to the canonical basis is 1 and is found in the second position of the 2-tuple. The first coordinate of $\binom{2}{1}$ relative to the basis $\mathbf{b}_1, \mathbf{b}_2$ is $\frac{3}{2}$ and the second coordinate of $\binom{2}{1}$ relative to the basis $\mathbf{b}_1, \mathbf{b}_2$ is $\frac{1}{2}$ and neither coordinate is a member of the 2-tuple in this case.

The reason this example was given in such great detail is that a notational convention which is used in this book is a possible source of great confusion. When it is understood that one is using a given ordered basis for a vector space of dimension n, it is often convenient to write a vector as the ordered n-tuple of its coordinates relative to the given ordered basis. The ith coordinate of a vector relative to the ordered basis occupies the ith position of the n-tuple. This coordinate representation will be used for all vectors in this text. Therefore, for example, the $\binom{3/2}{1/2}$ 2-tuple represents a unique vector only when the ordered basis or coordinate system is known, and the source of possible confusion is obvious if one forgets which ordered basis or coordinate system is being used. It should also be pointed out that to add two vectors, one must add the n-tuples of both vectors expressed relative to the same coordinate system. If one is using the canonical basis with its usual ordering, a vector in its usual representation and its coordinate representation are identical. Consider our previous example in which the coordinate representation of the $\binom{2}{1}$ vector relative to the canonical basis is $\binom{2}{1}$ but its coordinate representation relative to the $\mathbf{b}_1,\mathbf{b}_2$ basis is $\binom{3/2}{1/2}$. Therefore $\binom{2}{1}$ and $\binom{3/2}{1/2}$ can represent the same vector with respect to different bases. As long as the particular coordinate system being used is understood, there should be no confusion; also, we shall not use subscripts to denote the respective basis being used for the coordinate representation, but the basis being used will be clear from the context.

In the three-dimensional space of the physical world once the origin is specified, intuitively there is a one to one correspondence between all points in this space and all vectors which emanate from the origin to that point (or arrows whose tail is at the origin). In accordance with this intuitive picture, we may think of n-tuples either as points or as vectors from the origin to those points in n-dimensional space.

One is also used to picturing a rectangular coordinate system, that is, a set of coordinate axes at right angles using the same scale for each axis. This can be generalized from three to higher dimensions if we introduce the related notions of lengths and angles to a vector space. This is done through the notion of the inner product of two vectors.

DEFINITION 5 The (usual) *inner product* of

$$\mathbf{x} = \begin{pmatrix} x_1 \\ x_2 \\ \vdots \\ x_n \end{pmatrix} \quad \text{and} \quad \mathbf{y} = \begin{pmatrix} y_1 \\ y_2 \\ \vdots \\ y_n \end{pmatrix}$$

is $\sum_{i=1}^{n} x_i y_i$.

As a notational convenience we may think of our n-tuples as column vectors so that the transpose of $\mathbf{x}$ is the row vector $\mathbf{x}' = (x_1, \ldots, x_n)$ and we can write our inner product using matrix multiplication as $\mathbf{x}'\mathbf{y} = \sum_{i=1}^{n} x_i y_i$.

DEFINITION 6 The *length or norm* of a vector $\mathbf{x}$ is $\|\mathbf{x}\| = \sqrt{\mathbf{x}'\mathbf{x}}$.

DEFINITION 7 The *cosine of the angle between nonzero vectors* $\mathbf{x}$ *and* $\mathbf{y}$ is given by $\cos\theta_{\mathbf{x},\mathbf{y}} = \mathbf{x}'\mathbf{y}/\|\mathbf{x}\|\,\|\mathbf{y}\|$.

DEFINITION 8 Two vectors $\mathbf{x}$ and $\mathbf{y}$ are *orthogonal* or *perpendicular* if $\mathbf{x}'\mathbf{y} = 0$ (or $\cos\theta_{\mathbf{x},\mathbf{y}} = 0$).

THEOREM 5 Every orthogonal set of nonzero vectors is a linearly independent set.

DEFINITION 9 A set of mutually orthogonal vectors of unit length are *orthonormal.*

The inner product of two vectors is the same regardless of which coordinate system is used as long as the same coordinate system is used to represent both vectors.

THEOREM 6 Every (finite-dimensional) vector space with an inner product has an orthonormal basis.

THEOREM 7 If $\mathbf{b}_1, \ldots, \mathbf{b}_n$ is an orthonormal basis, then any vector $\mathbf{x}$ is uniquely expressed in terms of this basis as

$$\mathbf{x} = x_1\mathbf{b}_1 + \cdots + x_n\mathbf{b}_n \qquad \text{where} \quad x_i = \mathbf{x}'\mathbf{b}_i.$$

Therefore, ordered orthonormal bases are especially useful since the ith coordinate of a vector is the inner product of that vector with the ith ordered orthonormal basis vector. They are geometrically appealing since the ordered orthonormal basis or coordinate system is the higher-dimensional analogue of the rectangular coordinate system using the same scale for each axis in three dimensions.

In general, there are many different possible orthonormal bases for a given vector space. One should be perfectly willing to change the orthonormal basis or coordinate system if a more natural choice appears. The best technique is to select an orthonormal basis or orthonormal coordinate system which best fits the problem at hand. This is usually done on the basis of the geometry of the problem at hand.

If one decides to change bases, one would like a formula for the new coordinates of a vector in terms of the old coordinates. For ordered orthonormal coordinate systems, this is simple since the ith coordinate of a vector is the inner product of that vector with the ith ordered orthonormal basis vector.

THEOREM 8 Let $\boldsymbol{\delta}_1, \ldots, \boldsymbol{\delta}_n$ and $\mathbf{b}_1, \ldots, \mathbf{b}_n$ be respectively the old and new ordered orthonormal coordinate systems expressed as n-tuples with respect to the $\boldsymbol{\delta}_i$ coordinate system. If the vector $\mathbf{x}$ is expressed with respect to the $\boldsymbol{\delta}_i$ coordinate system, then this vector expressed with respect to the $\mathbf{b}_i$ coordinate system is the n-tuple

$$\begin{pmatrix} \mathbf{b}_1'\mathbf{x} \\ \vdots \\ \mathbf{b}_n'\mathbf{x} \end{pmatrix}.$$

DEFINITION 10 A linear transformation (or operator) $\mathbf{T}$ on a vector space V is a function that assigns to every vector $\mathbf{x}$ in V a vector $\mathbf{Tx}$ in V such that $\mathbf{T}(\alpha\mathbf{x} + \beta\mathbf{y}) = \alpha\mathbf{Tx} + \beta\mathbf{Ty}$ for all vectors $\mathbf{x}$ and $\mathbf{y}$ and scalars α and β.

While the idea of a linear transformation does not depend on a coordinate system, just as we can represent a vector by its coordinates relative to an ordered basis, we can represent a linear transformation by a square matrix relative to an ordered basis.

DEFINITION 11 Let $\mathbf{b}_1, \ldots, \mathbf{b}_n$ be an ordered basis for an n-dimensional vector space and let $\mathbf{T}$ be a linear transformation on this vector space. Since every vector is a linear combination of the $\mathbf{b}_i$, we can write

$$\mathbf{Tb}_j = \sum_{i=1}^{n} t_{ij}\mathbf{b}_i \qquad \text{for} \quad j = 1, \ldots, n.$$

The n^2 scalars t_{ij} are usually written in the form of a square matrix $\mathbf{T} = (t_{ij})$. This matrix is called the matrix of the linear transformation relative to the given ordered basis.

Note that in expressing $\mathbf{Tb}_j$ as a linear combination of $\mathbf{b}_1, \ldots, \mathbf{b}_n$, we write the coefficients so obtained as the jth column of the matrix $\mathbf{T}$. While it is usually customary to use the same symbol for the matrix as the transformation, if we change from one coordinate system to another, the same linear transformation may be represented by different matrices with respect to different bases and different symbols will be necessary. As usual, the basis or coordinate system being used will be clear from the context.

Recall that to add two linear transformations, one would add their corresponding matrices relative to the same basis. To multiply or compose two linear transformations, one would matrix multiply their corresponding matrices relative to the same basis (in the same order as they were composed).

The language of matrix algebra and of linear transformations are equivalent. Just as one talks about an invertible matrix, one can talk about an invertible transformation. Since, for example, a linear transformation is invertible if and only if the matrix corresponding to the transformation is invertible, we shall not review the definitions of many concepts about linear transformations as these correspond to the well-known matrix definitions.

Consider the n-dimensional vector space R^n; in the general case, an invertible linear transformation may be considered as a transformation of coordinates. However, we shall consider only orthonormal coordinate systems here, and a special type of invertible transformation suffices in this case.

THEOREM 9 Let $\boldsymbol{\delta}_1, \ldots, \boldsymbol{\delta}_n$ and $\mathbf{b}_1, \ldots, \mathbf{b}_n$ be respectively the old and new orthonormal coordinate systems expressed as n-tuples with respect to the $\boldsymbol{\delta}_i$ coordinate system. Let

$$\mathbf{x} = \begin{pmatrix} x_1 \\ \vdots \\ x_n \end{pmatrix} \quad \text{and} \quad \mathbf{z} = \begin{pmatrix} z_1 \\ \vdots \\ z_n \end{pmatrix}$$

be the column vectors in R^n of the old and new coordinates respectively of a given vector, and let the n by n matrix $\mathbf{P}$ be expressed in terms of its rows as

$$\mathbf{P} = \begin{pmatrix} \mathbf{b}_1' \\ \vdots \\ \mathbf{b}_n' \end{pmatrix}.$$

Then $\mathbf{z} = \mathbf{P}\mathbf{x}$. Also note that the matrix $\mathbf{P}$ is orthogonal; that is, $\mathbf{P}\mathbf{P}' = \mathbf{P}'\mathbf{P} = \mathbf{I}$.

For an n-dimensional vector space with an inner product, we may picture a set of orthogonal coordinate axes and a point in n space. The origin will always remain fixed. Consider changing orthonormal coordinate systems; we might view this as "erasing" our original coordinate axes and "redrawing" another more convenient set. This is equivalent to a rigid rotation or a rigid rotation followed by reflections of the n coordinate axes about their origin. In this case, the point remains fixed and we "move" our axes. Now consider the coordinate axes as fixed and move the point in such a fashion that distances are preserved. The linear transformations which preserve distances are the orthogonal transformations. We may now view an orthogonal linear transformation as equivalent to the process of changing from one orthonormal coordinate system to another.

If one decides to change bases, just as it is useful to have a formula for the new coordinates of a vector in terms of the old coordinates, it is useful to have a formula for the matrix of a linear transformation relative to the new basis in terms of the matrix of the transformation relative to the old basis. We shall do this only for ordered orthonormal coordinate systems.

THEOREM 10 Let $\boldsymbol{\delta}_1, \ldots, \boldsymbol{\delta}_n$ and $\mathbf{b}_1, \ldots, \mathbf{b}_n$ be respectively the old and new ordered orthonormal coordinate systems expressed as n-tuples with respect to the $\boldsymbol{\delta}_i$ coordinate system. Let $\mathbf{T}_\delta$ and $\mathbf{T}_b$ be respectively the matrix representations of a linear transformation relative to the $\boldsymbol{\delta}_i$ and $\mathbf{b}_i$ orthonormal coordinate systems, and let the n by n matrix $\mathbf{P}$ be expressed in terms of its rows as

$$\mathbf{P} = \begin{pmatrix} \mathbf{b}_1' \\ \vdots \\ \mathbf{b}_n' \end{pmatrix}.$$

Then $\mathbf{T}_b = \mathbf{P}\mathbf{T}_\delta\mathbf{P}'$. Note that the matrix $\mathbf{P}$ is orthogonal. (The matrix $\mathbf{P}'$ can be considered the matrix representation of a linear transformation $\mathbf{P}'$ relative to the $\boldsymbol{\delta}_i$ coordinate system, namely, the linear transformation defined by $\mathbf{P}'\boldsymbol{\delta}_i = \mathbf{b}_i$.)

One good reason for changing orthonormal coordinate systems is to represent a given linear transformation by a matrix with a particularly simple form. Perhaps the simplest nontrivial matrices to work with are the diagonal matrices, that is, those matrices whose off diagonal entries are zero. Recall that the rank of a diagonal matrix equals the number of nonzero diagonal entries and that the determinant of a diagonal matrix equals the product of the diagonal entries.

Given a linear transformation $\mathbf{T}$, is it possible to find an orthonormal basis $\mathbf{b}_1, \ldots, \mathbf{b}_n$ such that the matrix $\mathbf{D}$ of $\mathbf{T}$ relative to this basis is diagonal? Now

$$\mathbf{D} = \begin{pmatrix} \lambda_1 & 0 & \cdots & 0 \\ 0 & \lambda_2 & & 0 \\ \vdots & & \ddots & \vdots \\ 0 & 0 & \cdots & \lambda_n \end{pmatrix}$$

if and only if $\mathbf{T}\mathbf{b}_j = \lambda_j \mathbf{b}_j$, which suggests we should study vectors which are sent by a linear transformation into scalar multiples of themselves.

DEFINITION 12 A scalar λ is called an *eigenvalue* (characteristic root, latent root, proper value, or spectral value) and a nonzero vector $\mathbf{v}$ is called an *eigenvector* (characteristic vector, latent vector, proper vector, or spectral vector) of a linear transformation $\mathbf{T}$ if $\mathbf{T}\mathbf{v} = \lambda\mathbf{v}$.

THEOREM 11 A linear transformation on a vector space of dimension n has at most n distinct eigenvalues.

DEFINITION 13 A linear transformation $\mathbf{T}$ is diagonalizable if there is a basis consisting of eigenvectors.

DEFINITION 14 A linear transformation $\mathbf{T}$ on a vector space with the usual inner product is called a symmetric transformation if $(\mathbf{T}\mathbf{x})'\mathbf{y} = \mathbf{x}'(\mathbf{T}\mathbf{y})$ for all vectors $\mathbf{x}$ and $\mathbf{y}$.

THEOREM 12 A linear transformation is symmetric if and only if the matrix of the transformation with respect to an orthonormal basis is symmetric.

THEOREM 13 Let $\mathbf{T}$ be a symmetric linear transformation. Then there exists an orthonormal basis consisting of eigenvectors and hence symmetric linear transformations are diagonalizable.

THEOREM 14 (*Principal axis theorem*) If $\mathbf{C}$ is a real symmetric matrix (on R^n with the usual inner product), then there exists an orthogonal matrix $\mathbf{P}$ such that $\mathbf{PCP}' = \mathbf{D}$, where $\mathbf{D}$ is a diagonal matrix whose diagonal elements are the eigenvalues of $\mathbf{C}$ and the rows of $\mathbf{P}$ are normalized eigenvectors of $\mathbf{C}$ which are called the principal axes; that is, letting $\mathbf{D} = \text{diag}(d_i)$, the d_i are eigenvalues of $\mathbf{C}$, and $\mathbf{C}\mathbf{p}_i = d_i\mathbf{p}_i$, where $\mathbf{p}_i$ is the ith row of $\mathbf{P}$. Alternatively, we may write the matrix $\mathbf{C}$ as $\sum_{i=1}^{n} d_i\mathbf{p}_i\mathbf{p}_i'$.

DEFINITION 15 A vector space V is said to be the *direct sum* of the subspaces $S_1, \ldots, S_k$, denoted by $V = S_1 \oplus S_2 \oplus \cdots \oplus S_k$, if every vector $\mathbf{v}$ in V can be expressed uniquely in the form $\mathbf{v} = \mathbf{v}_1 + \mathbf{v}_2 + \cdots + \mathbf{v}_k$, where $\mathbf{v}_i \in S_i$ for $i = 1, \ldots, k$. If we drop the word unique from the definition, we say that V is the *sum* of the subspaces and denote this by $V = S_1 + S_2 + \cdots + S_k$.

Example Let V be a n-dimensional vector space with basis $\mathbf{b}_1, \ldots, \mathbf{b}_n$. If S_i is the one-dimensional subspace spanned by $\mathbf{b}_i$, $i = 1, \ldots, n$, then $V = S_1 \oplus S_2 \oplus \cdots \oplus S_n$.

The following theorem permits us to consider the direct sum of several subspaces as the result of a sequence of direct sums with only two subspaces.

THEOREM 15 If a summand in a direct sum decomposition of a vector space is replaced by a direct sum decomposition of that subspace, then a new direct sum decomposition of the entire vector space is obtained.

THEOREM 16 The sum of two subspaces is direct if and only if the intersection of these subspaces is the zero vector.

THEOREM 17 The dimension of a direct sum of subspaces is the sum of their dimensions.

THEOREM 18 If the dimension of a sum of subspaces is the sum of their dimensions, then the sum is direct.

DEFINITION 16 Subspaces $S_1, S_2, \ldots, S_k$ are *orthogonal* if any two vectors chosen from two distinct subspaces are orthogonal.

DEFINITION 17 A sum of subspaces is called an *orthogonal sum* if the subspaces are orthogonal.

THEOREM 19 An orthogonal sum of subspaces is a direct sum.

DEFINITION 18 A vector is orthogonal to a set of vectors S if it is orthogonal to each vector in S. The set of all vectors in a given vector space V which are orthogonal to S is called the *orthogonal complement* of S and is denoted by $S^{\perp}$.

Since a subspace of an inner product vector space may itself be considered as an inner product vector space, we should note that the vector space V in the previous definition may be a subspace of a larger vector space. The vector space relative to which the orthogonal complement is taken should be clear from the context.

THEOREM 20 The orthogonal complement of *any* nonempty set of vectors is a subspace.

THEOREM 21 Any vector space V with an inner product is the direct sum of any subspace S and its orthogonal complement, that is, $V = S \oplus S^{\perp}$.

This particular kind of direct sum decomposition (via a subspace and its orthogonal complement) is equivalent to the study of orthogonal projections. Since $V = S \oplus S^{\perp}$, every vector $\mathbf{v}$ in V can be expressed uniquely as $\mathbf{v} = \mathbf{x} + \mathbf{y}$ for $\mathbf{x} \in S$ and $\mathbf{y} \in S^{\perp}$. The vector $\mathbf{x}$ is called the *orthogonal projection* of the vector $\mathbf{v}$ onto the subspace S and the vector $\mathbf{y}$ is called the orthogonal projection of the vector $\mathbf{v}$ onto the subspace $S^{\perp}$. Since $S^{\perp}$ is uniquely determined by the subspace S, we need not specify both the direct summands but only one. The orthogonal projection of a vector $\mathbf{v}$ onto a subspace S is that vector in S which is geometrically "nearest" to $\mathbf{v}$. Or, equivalently, it is the "best" approximation to $\mathbf{v}$ by a vector in S in the sense of minimizing $\|\mathbf{v} - \mathbf{s}\|$ for $\mathbf{s} \in S$.

DEFINITION 19(a) The orthogonal projection operator onto S, a subspace of V, denoted by $\mathbf{P}_S$, is the transformation that assigns to each vector $\mathbf{v}$ in V its orthogonal projection on S, denoted by $\mathbf{P}_S\mathbf{v}$. Note that $\mathbf{P}_S\mathbf{v} \in S$.

THEOREM 22 The orthogonal projection operator is a linear transformation.

The orthogonal projection of a vector onto a subspace is given geometrically by dropping a perpendicular from the vector onto the subspace. Since a vector minus its orthogonal projection is orthogonal to S, we can use this property to give another equivalent definition of the orthogonal projection operator.

DEFINITION 19(b) The orthogonal projection operator onto S, a subspace of a vector space V, denoted by $\mathbf{P}_S$, assigns to each vector $\mathbf{v}$ in V the vector $\mathbf{P}_S\mathbf{v}$ in S such that $\mathbf{v} - \mathbf{P}_S\mathbf{v}$ is orthogonal to S.

Since we shall be dealing only with orthogonal projections, we will often drop the adjective orthogonal and just speak of projections. If the subspace we are projecting onto is clear from the context, we will often drop the subscript on the projection operator.

THEOREM 23 Let S be a subspace of V. The orthogonal projection of the vector $\mathbf{v}$ onto S has the same inner product as $\mathbf{v}$ does with each vector in S; that is, $(\mathbf{P}_S\mathbf{v})'\mathbf{y} = \mathbf{v}'\mathbf{y}$ for all $\mathbf{y} \in S$.

DEFINITION 20 A linear transformation $\mathbf{P}$ is *idempotent* if $\mathbf{P}\cdot\mathbf{P} = \mathbf{P}$; that is, $\mathbf{P}^2 = \mathbf{P}$.

THEOREM 24 Orthogonal projections are symmetric idempotent linear transformations.

Conversely, the properties of symmetry and idempotence characterize projection transformations.

THEOREM 25 Every symmetric idempotent transformation $\mathbf{P}$ is the orthogonal projection on the range of $\mathbf{P}$.

THEOREM 26 If $\mathbf{P}_S$ is the orthogonal projection onto a subspace S in a vector space V, then $\mathbf{I} - \mathbf{P}_S$ is the orthogonal projection onto $S^{\perp}$, where $\mathbf{I}$ is the identity transformation on V.

DEFINITION 21 Two projections $\mathbf{P}$ and $\mathbf{Q}$ are *orthogonal* if their composition is the zero transformation, that is, $\mathbf{P}\cdot\mathbf{Q} = \mathbf{0}$.

THEOREM 27 Projections $\mathbf{P}_S$ and $\mathbf{P}_T$ are orthogonal if and only if their corresponding subspaces S and T are orthogonal.

THEOREM 28 If S and T are subspaces of V, then $S \subseteq T$ if and only if $\mathbf{P}_T\cdot\mathbf{P}_S = \mathbf{P}_S$.

THEOREM 29 The sum of projections is a projection if and only if the projections are orthogonal. In this case, $\mathbf{P}_S + \mathbf{P}_T = \mathbf{P}_{S\oplus T}$.

THEOREM 30 If $S \subseteq T \subseteq V$, then $\mathbf{P}_T - \mathbf{P}_S$ is the orthogonal projection onto the orthogonal complement of S relative to T; that is, $S^{\perp} \cap T$.

Let $\mathbf{P}_S$ be the projection onto a subspace S. Note that if $\mathbf{v}$ is in S, then $\mathbf{P}_S\mathbf{v} = \mathbf{v}$, and if $\mathbf{v}$ is in $S^{\perp}$, then $\mathbf{P}_S\mathbf{v} = \mathbf{0}$. These are the only vectors projected into multiples of themselves.

THEOREM 31 The eigenvalues of a projection are zero or one.

THEOREM 32 If for some orthonormal coordinate system the matrix of a linear transformation $\mathbf{P}$ is diagonal with 0's and 1's along the diagonal, then $\mathbf{P}$ is a projection transformation.

DEFINITION 22 Let

$$\mathbf{x} = \begin{pmatrix} x_1 \\ \vdots \\ x_n \end{pmatrix}$$

be a vector variable. A *quadratic form* in $\mathbf{x}$ is a function

$$f(\mathbf{x}) = \sum_{i=1}^{n} \sum_{j=1}^{n} a_{ij} x_i x_j.$$

For any quadratic form, we may write $f(\mathbf{x}) = \mathbf{x}'\mathbf{Q}\mathbf{x}$, where $\mathbf{Q}$ is a symmetric n by n matrix. $\mathbf{Q}$ (uniquely determined) has elements $q_{ij} = (a_{ij} + a_{ji})/2$. $\mathbf{Q}$ is called the matrix of the quadratic form. We shall speak about both the quadratic form $\mathbf{Q}$ and the matrix $\mathbf{Q}$ (a slight abuse of terminology); for example, the quadratic form is called positive definite (nonnegative definite) if $\mathbf{Q}$ is positive definite (nonnegative definite).

DEFINITION 23 A matrix $\mathbf{C}$ is *positive definite* if $\mathbf{x}'\mathbf{C}\mathbf{x} > 0$ for all nonzero vectors $\mathbf{x}$.

DEFINITION 24 A matrix $\mathbf{C}$ is *nonnegative definite* if $\mathbf{x}'\mathbf{C}\mathbf{x} \geq 0$ for all vectors $\mathbf{x}$.

THEOREM 33 A real symmetric matrix $\mathbf{Q}$ is positive (nonnegative) definite if and only if all the eigenvalues are positive (nonnegative).

THEOREM 34 If a real symmetric matrix is positive definite, then it is nonsingular.

THEOREM 35 The inverse matrix of a positive definite matrix is positive definite.

THEOREM 36 If $\mathbf{Q}$ is a real symmetric positive definite matrix, then any matrix obtained from $\mathbf{Q}$ by deleting the same rows and columns of $\mathbf{Q}$ is positive definite. Therefore, the diagonal elements of a positive-definite matrix are positive.

APPENDIX 2 TABLES OF STATISTICAL DISTRIBUTIONS[1]

[1] Reprinted from R. Odeh, D. B. Owen, Z. W. Birnbaum, and L. Fisher, *Pocketbook of Statistical Tables*, Dekker, New York, 1977, by courtesy of Marcel Dekker, Inc.

TABLE 1

Critical Values for the t- and Normal Distributions

Let T_f be a t random variable with f df. Tabulated are values t_f such that $P(T_f \leq t_f) = \gamma$. Note that T_∞ is an $N(0, 1)$ variable.

	cumulative probability, γ						
f	0.70	0.80	0.90	0.95	0.975	0.990	0.995
1	0.727	1.376	3.0777	6.3138	12.7062	31.8205	63.6567
2	0.617	1.061	1.8856	2.9200	4.3027	6.9646	9.9248
3	0.584	0.978	1.6378	2.3534	3.1824	4.5407	5.8409
4	0.569	0.941	1.5332	2.1319	2.7766	3.7470	4.6041
5	0.559	0.920	1.4759	2.0151	2.5706	3.3651	4.0322
6	0.553	0.906	1.4398	1.9432	2.4469	3.1427	3.7075
7	0.549	0.896	1.4149	1.8946	2.3646	2.9980	3.4995
8	0.546	0.889	1.3968	1.8595	2.3060	2.8965	3.3554
9	0.543	0.883	1.3830	1.8331	2.2622	2.8214	3.2498
10	0.542	0.879	1.3722	1.8125	2.2281	2.7638	3.1693
11	0.540	0.876	1.3634	1.7959	2.2010	2.7181	3.1058
12	0.539	0.873	1.3562	1.7823	2.1788	2.6810	3.0545
13	0.538	0.870	1.3502	1.7709	2.1604	2.6503	3.0123
14	0.537	0.868	1.3450	1.7613	2.1448	2.6245	2.9768
15	0.536	0.866	1.3406	1.7531	2.1314	2.6025	2.9467
16	0.535	0.865	1.3368	1.7459	2.1199	2.5835	2.9208
17	0.534	0.863	1.3334	1.7396	2.1098	2.5669	2.8982
18	0.534	0.862	1.3304	1.7341	2.1009	2.5524	2.8784
19	0.533	0.861	1.3277	1.7291	2.0930	2.5395	2.8609
20	0.533	0.860	1.3253	1.7247	2.0860	2.5280	2.8453
21	0.532	0.859	1.3232	1.7207	2.0796	2.5176	2.8314
22	0.532	0.858	1.3212	1.7171	2.0739	2.5083	2.8188
23	0.532	0.858	1.3195	1.7139	2.0687	2.4999	2.8073
24	0.531	0.857	1.3178	1.7109	2.0639	2.4922	2.7969
25	0.531	0.856	1.3163	1.7081	2.0595	2.4851	2.7874
26	0.531	0.856	1.3150	1.7056	2.0555	2.4786	2.7787
27	0.531	0.855	1.3137	1.7033	2.0518	2.4727	2.7707
28	0.530	0.855	1.3125	1.7011	2.0484	2.4671	2.7633
29	0.530	0.854	1.3114	1.6991	2.0452	2.4620	2.7564
30	0.530	0.854	1.3104	1.6973	2.0423	2.4573	2.7500
31	0.530	0.853	1.3095	1.6955	2.0395	2.4528	2.7440
32	0.530	0.853	1.3086	1.6939	2.0369	2.4487	2.7385
33	0.530	0.853	1.3077	1.6924	2.0345	2.4448	2.7333
34	0.529	0.852	1.3070	1.6909	2.0322	2.4411	2.7284
35	0.529	0.852	1.3062	1.6896	2.0301	2.4377	2.7238
36	0.529	0.852	1.3055	1.6883	2.0281	2.4345	2.7195
37	0.529	0.851	1.3049	1.6871	2.0262	2.4314	2.7154
38	0.529	0.851	1.3042	1.6860	2.0244	2.4286	2.7116
39	0.529	0.851	1.3036	1.6849	2.0227	2.4258	2.7079
40	0.529	0.851	1.3031	1.6839	2.0211	2.4233	2.7045

TABLE 1 (continued)

	cumulative probability, γ						
f	0.70	0.80	0.90	0.95	0.975	0.990	0.995
42	0.528	0.850	1.3020	1.6820	2.0181	2.4185	2.6981
44	0.528	0.850	1.3011	1.6802	2.0154	2.4141	2.6923
46	0.528	0.850	1.3002	1.6787	2.0129	2.4102	2.6870
48	0.528	0.849	1.2994	1.6772	2.0106	2.4066	2.6822
50	0.528	0.849	1.2987	1.6759	2.0086	2.4033	2.6778
55	0.527	0.848	1.2971	1.6730	2.0040	2.3961	2.6682
60	0.527	0.848	1.2958	1.6706	2.0003	2.3901	2.6603
65	0.527	0.847	1.2947	1.6686	1.9971	2.3851	2.6536
70	0.527	0.847	1.2938	1.6669	1.9944	2.3808	2.6479
75	0.527	0.846	1.2929	1.6654	1.9921	2.3771	2.6430
80	0.526	0.846	1.2922	1.6641	1.9901	2.3739	2.6387
85	0.526	0.846	1.2916	1.6630	1.9883	2.3710	2.6349
90	0.526	0.846	1.2910	1.6620	1.9867	2.3685	2.6316
95	0.526	0.845	1.2905	1.6611	1.9853	2.3662	2.6286
100	0.526	0.845	1.2901	1.6602	1.9840	2.3642	2.6259
110	0.526	0.845	1.2893	1.6588	1.9818	2.3607	2.6213
120	0.526	0.845	1.2886	1.6577	1.9799	2.3578	2.6174
130	0.526	0.844	1.2881	1.6567	1.9784	2.3554	2.6142
140	0.526	0.844	1.2876	1.6558	1.9771	2.3533	2.6114
150	0.526	0.844	1.2872	1.6551	1.9759	2.3515	2.6090
160	0.525	0.844	1.2869	1.6544	1.9749	2.3499	2.6069
170	0.525	0.844	1.2866	1.6539	1.9740	2.3485	2.6051
180	0.525	0.844	1.2863	1.6534	1.9732	2.3472	2.6034
190	0.525	0.844	1.2860	1.6529	1.9725	2.3461	2.6020
200	0.525	0.843	1.2858	1.6525	1.9719	2.3451	2.6006
300	0.525	0.843	1.2844	1.6499	1.9679	2.3388	2.5923
400	0.525	0.843	1.2837	1.6487	1.9659	2.3357	2.5882
500	0.525	0.842	1.2832	1.6479	1.9647	2.3338	2.5857
600	0.525	0.842	1.2830	1.6474	1.9639	2.3326	2.5840
700	0.525	0.842	1.2828	1.6470	1.9634	2.3317	2.5829
800	0.525	0.842	1.2826	1.6468	1.9629	2.3310	2.5820
900	0.525	0.842	1.2825	1.6465	1.9626	2.3305	2.5813
1000	0.525	0.842	1.2824	1.6464	1.9623	2.3301	2.5808
∞	0.524	0.842	1.2816	1.6449	1.9600	2.3263	2.5758

TABLE 2

Critical Values for the F-Distribution

Let $F(f_1, f_2)$ be an F random variable with f_1 numerator degrees of freedom and f_2 denominator degrees of freedom. Tabulated for given α are critical values f such that $P(F(f_1, f_2) \geq f) = \alpha$.

α = .10

f_2 \ f_1	1	2	3	4	5	6	7	8	9
2	8.53	9.00	9.16	9.24	9.29	9.33	9.35	9.37	9.38
3	5.54	5.46	5.39	5.34	5.31	5.28	5.27	5.25	5.24
4	4.54	4.32	4.19	4.11	4.05	4.01	3.98	3.95	3.94
5	4.06	3.78	3.62	3.52	3.45	3.40	3.37	3.34	3.32
6	3.78	3.46	3.29	3.18	3.11	3.05	3.01	2.98	2.96
7	3.59	3.26	3.07	2.96	2.88	2.83	2.78	2.75	2.72
8	3.46	3.11	2.92	2.81	2.73	2.67	2.62	2.59	2.56
9	3.36	3.01	2.81	2.69	2.61	2.55	2.51	2.47	2.44
10	3.29	2.92	2.73	2.61	2.52	2.46	2.41	2.38	2.35
11	3.23	2.86	2.66	2.54	2.45	2.39	2.34	2.30	2.27
12	3.18	2.81	2.61	2.48	2.39	2.33	2.28	2.24	2.21
13	3.14	2.76	2.56	2.43	2.35	2.28	2.23	2.20	2.16
14	3.10	2.73	2.52	2.39	2.31	2.24	2.19	2.15	2.12
15	3.07	2.70	2.49	2.36	2.27	2.21	2.16	2.12	2.09
16	3.05	2.67	2.46	2.33	2.24	2.18	2.13	2.09	2.06
17	3.03	2.64	2.44	2.31	2.22	2.15	2.10	2.06	2.03
18	3.01	2.62	2.42	2.29	2.20	2.13	2.08	2.04	2.00
19	2.99	2.61	2.40	2.27	2.18	2.11	2.06	2.02	1.98
20	2.97	2.59	2.38	2.25	2.16	2.09	2.04	2.00	1.96
21	2.96	2.57	2.36	2.23	2.14	2.08	2.02	1.98	1.95
22	2.95	2.56	2.35	2.22	2.13	2.06	2.01	1.97	1.93
23	2.94	2.55	2.34	2.21	2.11	2.05	1.99	1.95	1.92
24	2.93	2.54	2.33	2.19	2.10	2.04	1.98	1.94	1.91
25	2.92	2.53	2.32	2.18	2.09	2.02	1.97	1.93	1.89
26	2.91	2.52	2.31	2.17	2.08	2.01	1.96	1.92	1.88
27	2.90	2.51	2.30	2.17	2.07	2.00	1.95	1.91	1.87
28	2.89	2.50	2.29	2.16	2.06	2.00	1.94	1.90	1.87
29	2.89	2.50	2.28	2.15	2.06	1.99	1.93	1.89	1.86
30	2.88	2.49	2.28	2.14	2.05	1.98	1.93	1.88	1.85
40	2.84	2.44	2.23	2.09	2.00	1.93	1.87	1.83	1.79
48	2.81	2.42	2.20	2.07	1.97	1.90	1.85	1.80	1.77
60	2.79	2.39	2.18	2.04	1.95	1.87	1.82	1.77	1.74
90	2.76	2.36	2.15	2.01	1.91	1.84	1.78	1.74	1.70
120	2.75	2.35	2.13	1.99	1.90	1.82	1.77	1.72	1.68
∞	2.71	2.30	2.08	1.94	1.85	1.77	1.72	1.67	1.63

TABLE 2 (continued)

$\alpha = .10$

f_2 \ f_1	10	12	15	20	24	30	40	60	120
2	9.39	9.41	9.42	9.44	9.45	9.46	9.47	9.47	9.48
3	5.23	5.22	5.20	5.18	5.18	5.17	5.16	5.15	5.14
4	3.92	3.90	3.87	3.84	3.83	3.82	3.80	3.79	3.78
5	3.30	3.27	3.24	3.21	3.19	3.17	3.16	3.14	3.12
6	2.94	2.90	2.87	2.84	2.82	2.80	2.78	2.76	2.74
7	2.70	2.67	2.63	2.59	2.58	2.56	2.54	2.51	2.49
8	2.54	2.50	2.46	2.42	2.40	2.38	2.36	2.34	2.32
9	2.42	2.38	2.34	2.30	2.28	2.25	2.23	2.21	2.18
10	2.32	2.28	2.24	2.20	2.18	2.16	2.13	2.11	2.08
11	2.25	2.21	2.17	2.12	2.10	2.08	2.05	2.03	2.00
12	2.19	2.15	2.10	2.06	2.04	2.01	1.99	1.96	1.93
13	2.14	2.10	2.05	2.01	1.98	1.96	1.93	1.90	1.88
14	2.10	2.05	2.01	1.96	1.94	1.91	1.89	1.86	1.83
15	2.06	2.02	1.97	1.92	1.90	1.87	1.85	1.82	1.79
16	2.03	1.99	1.94	1.89	1.87	1.84	1.81	1.78	1.75
17	2.00	1.96	1.91	1.86	1.84	1.81	1.78	1.75	1.72
18	1.98	1.93	1.89	1.84	1.81	1.78	1.75	1.72	1.69
19	1.96	1.91	1.86	1.81	1.79	1.76	1.73	1.70	1.67
20	1.94	1.89	1.84	1.79	1.77	1.74	1.71	1.68	1.64
21	1.92	1.87	1.83	1.78	1.75	1.72	1.69	1.66	1.62
22	1.90	1.86	1.81	1.76	1.73	1.70	1.67	1.64	1.60
23	1.89	1.84	1.80	1.74	1.72	1.69	1.66	1.62	1.59
24	1.88	1.83	1.78	1.73	1.70	1.67	1.64	1.61	1.57
25	1.87	1.82	1.77	1.72	1.69	1.66	1.63	1.59	1.56
26	1.86	1.81	1.76	1.71	1.68	1.65	1.61	1.58	1.54
27	1.85	1.80	1.75	1.70	1.67	1.64	1.60	1.57	1.53
28	1.84	1.79	1.74	1.69	1.66	1.63	1.59	1.56	1.52
29	1.83	1.78	1.73	1.68	1.65	1.62	1.58	1.55	1.51
30	1.82	1.77	1.72	1.67	1.64	1.61	1.57	1.54	1.50
40	1.76	1.71	1.66	1.61	1.57	1.54	1.51	1.47	1.42
48	1.73	1.69	1.63	1.57	1.54	1.51	1.47	1.43	1.39
60	1.71	1.66	1.60	1.54	1.51	1.48	1.44	1.40	1.35
90	1.67	1.62	1.56	1.50	1.47	1.43	1.39	1.35	1.29
120	1.65	1.60	1.55	1.48	1.45	1.41	1.37	1.32	1.26
∞	1.60	1.55	1.49	1.42	1.38	1.34	1.30	1.24	1.17

TABLE 2 (continued)

$\alpha = .05$

f_2 \ f_1	1	2	3	4	5	6	7	8	9
2	18.51	19.00	19.16	19.25	19.30	19.33	19.35	19.37	19.38
3	10.13	9.55	9.28	9.12	9.01	8.94	8.89	8.85	8.81
4	7.71	6.94	6.59	6.39	6.26	6.16	6.09	6.04	6.00
5	6.61	5.79	5.41	5.19	5.05	4.95	4.88	4.82	4.77
6	5.99	5.14	4.76	4.53	4.39	4.28	4.21	4.15	4.10
7	5.59	4.74	4.35	4.12	3.97	3.87	3.79	3.73	3.68
8	5.32	4.46	4.07	3.84	3.69	3.58	3.50	3.44	3.39
9	5.12	4.26	3.86	3.63	3.48	3.37	3.29	3.23	3.18
10	4.96	4.10	3.71	3.48	3.33	3.22	3.14	3.07	3.02
11	4.84	3.98	3.59	3.36	3.20	3.09	3.01	2.95	2.90
12	4.75	3.89	3.49	3.26	3.11	3.00	2.91	2.85	2.80
13	4.67	3.81	3.41	3.18	3.03	2.92	2.83	2.77	2.71
14	4.60	3.74	3.34	3.11	2.96	2.85	2.76	2.70	2.65
15	4.54	3.68	3.29	3.06	2.90	2.79	2.71	2.64	2.59
16	4.49	3.63	3.24	3.01	2.85	2.74	2.66	2.59	2.54
17	4.45	3.59	3.20	2.96	2.81	2.70	2.61	2.55	2.49
18	4.41	3.55	3.16	2.93	2.77	2.66	2.58	2.51	2.46
19	4.38	3.52	3.13	2.90	2.74	2.63	2.54	2.48	2.42
20	4.35	3.49	3.10	2.87	2.71	2.60	2.51	2.45	2.39
21	4.32	3.47	3.07	2.84	2.68	2.57	2.49	2.42	2.37
22	4.30	3.44	3.05	2.82	2.66	2.55	2.46	2.40	2.34
23	4.28	3.42	3.03	2.80	2.64	2.53	2.44	2.37	2.32
24	4.26	3.40	3.01	2.78	2.62	2.51	2.42	2.36	2.30
25	4.24	3.39	2.99	2.76	2.60	2.49	2.40	2.34	2.28
26	4.23	3.37	2.98	2.74	2.59	2.47	2.39	2.32	2.27
27	4.21	3.35	2.96	2.73	2.57	2.46	2.37	2.31	2.25
28	4.20	3.34	2.95	2.71	2.56	2.45	2.36	2.29	2.24
29	4.18	3.33	2.93	2.70	2.55	2.43	2.35	2.28	2.22
30	4.17	3.32	2.92	2.69	2.53	2.42	2.33	2.27	2.21
40	4.08	3.23	2.84	2.61	2.45	2.34	2.25	2.18	2.12
48	4.04	3.19	2.80	2.57	2.41	2.29	2.21	2.14	2.08
60	4.00	3.15	2.76	2.53	2.37	2.25	2.17	2.10	2.04
90	3.95	3.10	2.71	2.47	2.32	2.20	2.11	2.04	1.99
120	3.92	3.07	2.68	2.45	2.29	2.18	2.09	2.02	1.96
∞	3.84	3.00	2.60	2.37	2.21	2.10	2.01	1.94	1.88

TABLE 2 (continued)

α = .05

f_2 \ f_1	10	12	15	20	24	30	40	60	120
2	19.40	19.41	19.43	19.45	19.45	19.46	19.47	19.48	19.49
3	8.79	8.74	8.70	8.66	8.64	8.62	8.59	8.57	8.55
4	5.96	5.91	5.86	5.80	5.77	5.75	5.72	5.69	5.66
5	4.74	4.68	4.62	4.56	4.53	4.50	4.46	4.43	4.40
6	4.06	4.00	3.94	3.87	3.84	3.81	3.77	3.74	3.70
7	3.64	3.57	3.51	3.44	3.41	3.38	3.34	3.30	3.27
8	3.35	3.28	3.22	3.15	3.12	3.08	3.04	3.01	2.97
9	3.14	3.07	3.01	2.94	2.90	2.86	2.83	2.79	2.75
10	2.98	2.91	2.85	2.77	2.74	2.70	2.66	2.62	2.58
11	2.85	2.79	2.72	2.65	2.61	2.57	2.53	2.49	2.45
12	2.75	2.69	2.62	2.54	2.51	2.47	2.43	2.38	2.34
13	2.67	2.60	2.53	2.46	2.42	2.38	2.34	2.30	2.25
14	2.60	2.53	2.46	2.39	2.35	2.31	2.27	2.22	2.18
15	2.54	2.48	2.40	2.33	2.29	2.25	2.20	2.16	2.11
16	2.49	2.42	2.35	2.28	2.24	2.19	2.15	2.11	2.06
17	2.45	2.38	2.31	2.23	2.19	2.15	2.10	2.06	2.01
18	2.41	2.34	2.27	2.19	2.15	2.11	2.06	2.02	1.97
19	2.38	2.31	2.23	2.16	2.11	2.07	2.03	1.98	1.93
20	2.35	2.28	2.20	2.12	2.08	2.04	1.99	1.95	1.90
21	2.32	2.25	2.18	2.10	2.05	2.01	1.96	1.92	1.87
22	2.30	2.23	2.15	2.07	2.03	1.98	1.94	1.89	1.84
23	2.27	2.20	2.13	2.05	2.01	1.96	1.91	1.86	1.81
24	2.25	2.18	2.11	2.03	1.98	1.94	1.89	1.84	1.79
25	2.24	2.16	2.09	2.01	1.96	1.92	1.87	1.82	1.77
26	2.22	2.15	2.07	1.99	1.95	1.90	1.85	1.80	1.75
27	2.20	2.13	2.06	1.97	1.93	1.88	1.84	1.79	1.73
28	2.19	2.12	2.04	1.96	1.91	1.87	1.82	1.77	1.71
29	2.18	2.10	2.03	1.94	1.90	1.85	1.81	1.75	1.70
30	2.16	2.09	2.01	1.93	1.89	1.84	1.79	1.74	1.68
40	2.08	2.00	1.92	1.84	1.79	1.74	1.69	1.64	1.58
48	2.03	1.96	1.88	1.79	1.75	1.70	1.64	1.59	1.52
60	1.99	1.92	1.84	1.75	1.70	1.65	1.59	1.53	1.47
90	1.94	1.86	1.78	1.69	1.64	1.59	1.53	1.46	1.39
120	1.91	1.83	1.75	1.66	1.61	1.55	1.50	1.43	1.35
∞	1.83	1.75	1.67	1.57	1.52	1.46	1.39	1.32	1.22

TABLE 2 (continued)

α = .01

f_2 \ f_1	1	2	3	4	5	6	7	8	9
2	98.50	99.00	99.17	99.25	99.30	99.33	99.36	99.37	99.39
3	34.12	30.82	29.46	28.71	28.24	27.91	27.67	27.49	27.35
4	21.20	18.00	16.69	15.98	15.52	15.21	14.98	14.80	14.66
5	16.26	13.27	12.06	11.39	10.97	10.67	10.46	10.29	10.16
6	13.75	10.92	9.78	9.15	8.75	8.47	8.26	8.10	7.98
7	12.25	9.55	8.45	7.85	7.46	7.19	6.99	6.84	6.72
8	11.26	8.65	7.59	7.01	6.63	6.37	6.18	6.03	5.91
9	10.56	8.02	6.99	6.42	6.06	5.80	5.61	5.47	5.35
10	10.04	7.56	6.55	5.99	5.64	5.39	5.20	5.06	4.94
11	9.65	7.21	6.22	5.67	5.32	5.07	4.89	4.74	4.63
12	9.33	6.93	5.95	5.41	5.06	4.82	4.64	4.50	4.39
13	9.07	6.70	5.74	5.21	4.86	4.62	4.44	4.30	4.19
14	8.86	6.51	5.56	5.04	4.69	4.46	4.28	4.14	4.03
15	8.68	6.36	5.42	4.89	4.56	4.32	4.14	4.00	3.89
16	8.53	6.23	5.29	4.77	4.44	4.20	4.03	3.89	3.78
17	8.40	6.11	5.18	4.67	4.34	4.10	3.93	3.79	3.68
18	8.29	6.01	5.09	4.58	4.25	4.01	3.84	3.71	3.60
19	8.18	5.93	5.01	4.50	4.17	3.94	3.77	3.63	3.52
20	8.10	5.85	4.94	4.43	4.10	3.87	3.70	3.56	3.46
21	8.02	5.78	4.87	4.37	4.04	3.81	3.64	3.51	3.40
22	7.95	5.72	4.82	4.31	3.99	3.76	3.59	3.45	3.35
23	7.88	5.66	4.76	4.26	3.94	3.71	3.54	3.41	3.30
24	7.82	5.61	4.72	4.22	3.90	3.67	3.50	3.36	3.26
25	7.77	5.57	4.68	4.18	3.85	3.63	3.46	3.32	3.22
26	7.72	5.53	4.64	4.14	3.82	3.59	3.42	3.29	3.18
27	7.68	5.49	4.60	4.11	3.78	3.56	3.39	3.26	3.15
28	7.64	5.45	4.57	4.07	3.75	3.53	3.36	3.23	3.12
29	7.60	5.42	4.54	4.04	3.73	3.50	3.33	3.20	3.09
30	7.56	5.39	4.51	4.02	3.70	3.47	3.30	3.17	3.07
40	7.31	5.18	4.31	3.83	3.51	3.29	3.12	2.99	2.89
48	7.19	5.08	4.22	3.74	3.43	3.20	3.04	2.91	2.80
60	7.08	4.98	4.13	3.65	3.34	3.12	2.95	2.82	2.72
90	6.93	4.85	4.01	3.53	3.23	3.01	2.84	2.72	2.61
120	6.85	4.79	3.95	3.48	3.17	2.96	2.79	2.66	2.56
∞	6.63	4.61	3.78	3.32	3.02	2.80	2.64	2.51	2.41

TABLE 2 (continued)

α = .01

f_2 \ f_1	10	12	15	20	24	30	40	60	120
2	99.40	99.42	99.43	99.45	99.46	99.47	99.47	99.48	99.49
3	27.23	27.05	26.87	26.69	26.60	26.50	26.41	26.32	26.22
4	14.55	14.37	14.20	14.02	13.93	13.84	13.75	13.65	13.56
5	10.05	9.89	9.72	9.55	9.47	9.38	9.29	9.20	9.11
6	7.87	7.72	7.56	7.40	7.31	7.23	7.14	7.06	6.97
7	6.62	6.47	6.31	6.16	6.07	5.99	5.91	5.82	5.74
8	5.81	5.67	5.52	5.36	5.28	5.20	5.12	5.03	4.95
9	5.26	5.11	4.96	4.81	4.73	4.65	4.57	4.48	4.40
10	4.85	4.71	4.56	4.41	4.33	4.25	4.17	4.08	4.00
11	4.54	4.40	4.25	4.10	4.02	3.94	3.86	3.78	3.69
12	4.30	4.16	4.01	3.86	3.78	3.70	3.62	3.54	3.45
13	4.10	3.96	3.82	3.66	3.59	3.31	3.43	3.34	3.25
14	3.94	3.80	3.66	3.51	3.43	3.35	3.27	3.18	3.09
15	3.80	3.67	3.52	3.37	3.29	3.21	3.13	3.05	2.96
16	3.69	3.55	3.41	3.26	3.18	3.10	3.02	2.93	2.84
17	3.59	3.46	3.31	3.16	3.08	3.00	2.92	2.83	2.75
18	3.51	3.37	3.23	3.08	3.00	2.92	2.84	2.75	2.66
19	3.43	3.30	3.15	3.00	2.92	2.84	2.76	2.67	2.58
20	3.37	3.23	3.09	2.94	2.86	2.78	2.69	2.61	2.52
21	3.31	3.17	3.03	2.88	2.80	2.72	2.64	2.55	2.46
22	3.26	3.12	2.98	2.83	2.75	2.67	2.58	2.50	2.40
23	3.21	3.07	2.93	2.78	2.70	2.62	2.54	2.45	2.35
24	3.17	3.03	2.89	2.74	2.66	2.58	2.49	2.40	2.31
25	3.13	2.99	2.85	2.70	2.62	2.54	2.45	2.36	2.27
26	3.09	2.96	2.81	2.66	2.58	2.50	2.42	2.33	2.23
27	3.06	2.93	2.78	2.63	2.55	2.47	2.38	2.29	2.20
28	3.03	2.90	2.75	2.60	2.52	2.44	2.35	2.26	2.17
29	3.00	2.87	2.73	2.57	2.49	2.41	2.33	2.23	2.14
30	2.98	2.84	2.70	2.55	2.47	2.39	2.30	2.21	2.11
40	2.80	2.66	2.52	2.37	2.29	2.20	2.11	2.02	1.92
48	2.71	2.58	2.44	2.28	2.20	2.12	2.02	1.93	1.82
60	2.63	2.50	2.35	2.20	2.12	2.13	1.94	1.84	1.73
90	2.52	2.39	2.24	2.09	2.00	1.92	1.82	1.72	1.60
120	2.47	2.34	2.19	2.03	1.95	1.86	1.76	1.66	1.53
∞	2.32	2.18	2.04	1.89	1.79	1.70	1.59	1.47	1.32

TABLE 3

Critical Values of the Studentized Range from a Normal Distribution

Let $Q_{n,v}$ be as defined on page 106. Tabulated are values q such that for given n, v, and γ, $P(Q_{n,v} \le q) = \gamma$.

cumulative probability, γ = 0.90

ν	n=3	n=4	n=5	n=6	n=7	n=8	n=9	n=10
2	5.73	6.77	7.54	8.14	8.63	9.05	9.41	9.72
4	3.98	4.59	5.03	5.39	5.68	5.93	6.14	6.33
6	3.56	4.07	4.44	4.73	4.97	5.17	5.34	5.50
8	3.37	3.83	4.17	4.43	4.65	4.83	4.99	5.13
10	3.27	3.70	4.02	4.26	4.47	4.64	4.78	4.91
12	3.20	3.62	3.92	4.16	4.35	4.51	4.65	4.78
14	3.16	3.56	3.85	4.08	4.27	4.42	4.56	4.68
16	3.12	3.52	3.80	4.03	4.21	4.36	4.49	4.61
18	3.10	3.49	3.77	3.98	4.16	4.31	4.44	4.55
20	3.08	3.46	3.74	3.95	4.12	4.27	4.40	4.51
24	3.05	3.42	3.69	3.90	4.07	4.21	4.34	4.44
30	3.02	3.39	3.65	3.85	4.02	4.16	4.28	4.38
40	2.99	3.35	3.60	3.80	3.96	4.10	4.21	4.32
60	2.97	3.31	3.56	3.75	3.91	4.04	4.16	4.25
120	2.93	3.28	3.52	3.71	3.86	3.99	4.10	4.19
∞	2.90	3.24	3.48	3.66	3.81	3.93	4.04	4.13

cumulative probability, γ = 0.95

	n=3	n=4	n=5	n=6	n=7	n=8	n=9	n=10
2	8.33	9.80	10.88	11.73	12.43	13.03	13.54	13.99
4	5.04	5.76	6.29	6.71	7.05	7.35	7.60	7.83
6	4.34	4.90	5.30	5.63	5.90	6.12	6.32	6.49
8	4.04	4.53	4.89	5.17	5.40	5.60	5.77	5.92
10	3.88	4.33	4.65	4.91	5.12	5.30	5.46	5.60
12	3.77	4.20	4.51	4.75	4.95	5.12	5.27	5.39
14	3.70	4.11	4.41	4.64	4.83	4.99	5.13	5.25
16	3.65	4.05	4.33	4.56	4.74	4.90	5.03	5.15
18	3.61	4.00	4.28	4.49	4.67	4.82	4.96	5.07
20	3.58	3.96	4.23	4.45	4.62	4.77	4.90	5.01
24	3.53	3.90	4.17	4.37	4.54	4.68	4.81	4.92
30	3.49	3.85	4.10	4.30	4.46	4.60	4.72	4.82
40	3.44	3.79	4.04	4.23	4.39	4.52	4.63	4.73
60	3.40	3.74	3.98	4.16	4.31	4.44	4.55	4.65
120	3.36	3.68	3.92	4.10	4.24	4.36	4.47	4.56
∞	3.31	3.63	3.86	4.03	4.17	4.29	4.39	4.47

TABLE 3 (continued)

cumulative probability, γ = 0.975

ν	n=3	n=4	n=5	n=6	n=7	n=8	n=9	n=10
2	11.94	14.01	15.54	16.75	17.74	18.58	19.31	19.95
4	6.24	7.09	7.72	8.21	8.62	8.98	9.28	9.55
6	5.16	5.77	6.23	6.59	6.88	7.14	7.36	7.55
8	4.71	5.23	5.62	5.92	6.17	6.38	6.57	6.73
10	4.47	4.94	5.29	5.56	5.78	5.97	6.14	6.28
12	4.32	4.76	5.08	5.33	5.54	5.72	5.87	6.00
14	4.22	4.64	4.94	5.18	5.37	5.54	5.68	5.81
16	4.15	4.55	4.84	5.07	5.25	5.41	5.55	5.67
18	4.09	4.48	4.76	4.98	5.16	5.31	5.45	5.57
20	4.05	4.43	4.70	4.91	5.09	5.24	5.37	5.48
24	3.98	4.35	4.61	4.82	4.99	5.13	5.25	5.36
30	3.92	4.27	4.52	4.72	4.88	5.02	5.13	5.24
40	3.86	4.20	4.44	4.63	4.78	4.91	5.02	5.12
60	3.80	4.12	4.36	4.54	4.68	4.81	4.91	5.01
120	3.74	4.05	4.28	4.45	4.59	4.70	4.81	4.89
∞	3.68	3.98	4.20	4.36	4.49	4.60	4.70	4.78

cumulative probability, γ = 0.99

ν	n=3	n=4	n=5	n=6	n=7	n=8	n=9	n=10
2	19.02	22.29	24.72	26.63	28.20	29.53	30.68	31.69
4	8.12	9.17	9.96	10.58	11.10	11.54	11.93	12.26
6	6.33	7.03	7.56	7.97	8.32	8.61	8.87	9.10
8	5.64	6.20	6.62	6.96	7.24	7.47	7.68	7.86
10	5.27	5.77	6.14	6.43	6.67	6.87	7.05	7.21
12	5.05	5.50	5.84	6.10	6.32	6.51	6.67	6.81
14	4.89	5.32	5.63	5.88	6.08	6.26	6.41	6.54
16	4.79	5.19	5.49	5.72	5.92	6.08	6.22	6.35
18	4.70	5.09	5.38	5.60	5.79	5.94	6.08	6.20
20	4.64	5.02	5.29	5.51	5.69	5.84	5.97	6.09
24	4.55	4.91	5.17	5.37	5.54	5.69	5.81	5.92
30	4.45	4.80	5.05	5.24	5.40	5.54	5.65	5.76
40	4.37	4.70	4.93	5.11	5.26	5.39	5.50	5.60
60	4.28	4.59	4.82	4.99	5.13	5.25	5.36	5.45
120	4.20	4.50	4.71	4.87	5.01	5.12	5.21	5.30
∞	4.12	4.40	4.60	4.76	4.88	4.99	5.08	5.16

INDEX

D

E

S

T

U

V

W

A
B
C 8
D 9
E 0
F 1
G 2
H 3
I 4
J 5